EVOLUTION TO COMPLEXITY.

From unanimated matter to the universal superorganism

Edwin Francisco Herrera Paz, MD

ISBN: 1518628834
ISBN-13: 978-1518628832

DEDICATION

To all those who seek the truth beyond the
established paradigms.

INDEX

INTRODUCTION. COMPLEXITY AND LIFE

"Life is really simple, but we insist on making it complicated."

— Confucius

When trying to define what life is, we oftentimes meet up with the conundrum of its complexity. Defining life is most certainly a puzzling task, one which gets stressed even further when we attempt at differentiating between what is alive and what is not. We can certainly agree, even without being extremely knowledgeable about biological phenomena, that a sedimentary rock is not alive, while a bird flying across the sky is. While we find it obvious that the beds in which we sleep in are lifeless bodies, we are prompt to recognize that our house pets are live creatures. The prime distinction between these items seems quite simple, but then and all of a sudden complications arise once we approach the defining limits of what we call Life.

Viruses are perhaps the oldest evolutionary examples, relics of a prebiotic world. Any given virus has only one type of nucleic acid, either RNA or DNA, unlike the rest of all known living creatures, which have both. Viruses are unable to multiply by themselves so to achieve this they must hijack the machinery of another living organism in order to make it produce clones of themselves. Their structure is quite simple: Viruses are molecular machines composed of a set of proteins and nucleic acid. But then the question arises: Are they or are they not alive?

In giving another example: Platelets found in the bloodstream are the structures responsible for coagulation of blood in hemorrhages. They are a part of the solid (cellular) fraction of our blood, and that of other living species. Within this fraction we also find white blood cells, which are responsible for the immune activity, and red blood cells or erythrocytes whose main function is to carry oxygen from the lungs to the peripheral tissues. These three structures (platelets, leukocytes and erythrocytes) all originate from a progenitor stem cell found in the bone marrow. To give rise to platelets, a normal stem cell is modified, transforming into another cell called megakaryocyte. The latter is fragmented into several pieces and each one of them will become a platelet. Therefore platelets are not cells but cell fragments. So then again, the question arises: Are platelets living organisms?

We may believe this type of organisms stand at the limit of our definition of life, but that is not so. Viruses, evolutionarily, needed a large number of intermediate steps starting from inert matter before becoming what they are. It is not difficult to put platelets and viruses in the same sack of living beings. However, we can continue down the evolutionary ladder to the first self-replicating complex molecules. So then, are the proteins and other biological molecules living organisms? Of course not, you would say, but if we could reduce our size in order to spend a day in an average cell, living in the small universe of proteins, perhaps our concept of what is alive would be shifted. The activity within a cell is intense, with a huge variability of proteins performing diverse types of work and cooperating with each other like workers in a factory. If we could live in this world for a while we would wonder what stops us from including these "beings" in the category of the living. Perhaps we would then feel inclined to change the parameters on which we stand to define what an agent must have to be considered alive.

So now, what is really wrong with our categorization of what is alive and what is not? The answer falls onto an issue of "language restriction". We use the term life, at least in our realm, for a large number of organisms related by the same carbon chemistry and a common way of encoding information (the genetic code). "Life" includes organisms as diverse as an oak, a worm, a bacterium or a human, without distinction. You would not say that one of your skin cells from the tip of a toe is less alive than you are, considering "yourself" as a whole. Both are living things. But our psyche makes important distinctions between both which have significant conceptual and moral implications. For example, you will not instinctively find it a crime to end the life of one body cell, (of which we own approximately 15 trillion) but on the other hand you would likely consider a human life to be extremely valuable.

It becomes obvious that the duality between living/non-living organisms is insufficient to cover the broad spectrum from inanimate particles to intelligent organisms and to that of complex human communities or social insects. How then do we fill the gap in this concept? How can we cover our intuitive need to categorize some organisms as "more alive" than others? Because you cannot change the qualitative characteristic inherent to the word "life" to make it quantitative, we can make use of a word which enables us to save this situation. This term is "complexity".

Complexity is not just another trivial word, as suggested by everyday use. When we experience difficulties solving mathematical problems we say they are complex. The complexity of mathematical problems is derived from the amount of effort it takes to reach a solution, so, an appropriate synonym in everyday life would be "complicated". However, in the fields of science, complex and complicated are not the same. Complexity is a term applied to certain types of non-linear systems, formed by a variety of interconnected elements whose links create additional information, unseen before by the observer. As a result of interactions, new properties that cannot be explained by the properties of the individual elements emerge: the so-called emergent properties. It is possible to analyze the degree of complexity of a system based on a variety of parameters, but I will here mention three, perhaps the most important. In ascending order: 1) The number of elements of the system, 2) The mean number of relationships between elements, and 3) structural levels of the system.

Take for example the process of communication between people. Two persons talking form a system; however we have only two elements, so its complexity degree is low. In comparison, a meeting of the board of directors of a company is a more complex system that has a larger number of elements. A diplomatic cocktail where there are small groups of people chatting has even many more elements, but the number of relationships between each of these elements is still low. An Internet forum, however, draws on the views of many. —Here, every user can read and interact with every other user in the forum. In the latter case the number of relationships of participants is higher, increasing the degree of complexity. In fact, in order to determine the complexity of a system, the number of relationships between the elements is even more important than the number of elements.

Human communication systems referred to above are relatively simple since their structure is flat, without levels. A flat network seems very simple when compared to a multilevel structure (to understand the complexity of networks of relationships see Gleiser and Spoormaker, 2010; Apicella et al., 2012; Fowler, 2012). Within simple, flat networks every element is at the same level, and groups of items don't form new elements. A hierarchical system with several structural levels will be more complex than a horizontal one. The flow of information in a corporation will be hierarchical with multiple levels of command, from the workers, to middle management, up to top managers. We will call these "levels of complexity". The greater the number of levels, the more complex, and even more so when we add control mechanisms between and within levels that help maintain stability and good performance. A group of items at a particular level can be grouped to form in turn a functional unit with new emergent properties. Each of these units then becomes one element in a higher level. In our example of the corporation, employees (elements of the first level of complexity) can be organized into departments. Each department with all its employees is then an element from a higher level, which is called a corporation. In another example, our cells (elements of the first level) are organized to form specific organs, which are new elements of a higher level: a human body.

It is convenient at this time to clarify that the principles governing complex systems are not limited to living beings, but they are applied to a large number of phenomena in nature. Global climate, for example, is a complex system. However, for the purpose of this essay I will analyze mainly those special, singular and striking complex systems that deserve to be called "alive", as well as their derivatives, such as the social, political, economic, commercial, cultural, linguistic, and so on. Furthermore, biological complex systems and their derivatives differ from inanimate in a special feature: their enormous capacity to evolve towards higher levels of complexity.

Let's do an exercise and ask ourselves a question. How many levels of complexity are there in a transnational corporation? Various complex molecules called proteins cooperate to perform a task, forming a molecular motor or gear. We will establish the molecular level as the first. Molecular gears are bound structures with specific functions within the cell. For example: In the forming of organelles. This will be our second level. The organelles with different functions interact and cooperate to form a functional unit, the cell, which is also our third level. Various cell types interact to form a specific tissue, our fourth level. Tissues make up organs, the fifth level. Various organs cooperate to form a system, the sixth level. Systems are assembled to form a human being, the seventh level. Humans cooperate to form a department, the eighth level. The departments are combined to form a branch, the ninth level. Several branches in a country form a division, the tenth level, and several divisions in different countries form a transnational corporation, our eleventh level.

Eleven levels of complexity in a transnational corporation! And that is oversimplifying for illustrative purposes, not to mention that a corporation is not made only of human beings. In reality, living complex systems interact in complex ways as well. A human being consists not only of self tissues. In fact, the amount of cells of our own is less than the number of bacteria contained in the bacterial flora of our skin and gut (Turnbaugh et al., 2007; Mitreva, 2012). These bacteria form communities and ecosystems, with their own levels of complexity, but they are fully integrated into our bodies. A city is made up of citizens, but also by transportation, infrastructure, plant and animal species that serve as food, etc. The system consists of a number of basic elements that relate with each other to form structures, but also has a large number of alien elements. A corporation is not made only of humans. It may likely have computers, robots, buildings, vehicles, etc. Each individual element has its own history, its own implicit complexity, yet it can be part of multiple complex systems. An individual can become part of a transnational corporation, but also a citizen of a country, which is a complex system in its own right. The bacteria that live in our large intestine are part of an intricate micro ecosystem, but are also part of us.

When and where this cycle of creation of complexity did begin? Well, very far back in time; to be precise, about 13.7 billion years ago.

LITERATURE CITED

Apicella CL, Marlowe FW, Fowler JH, Christakis NA (2012). Social networks and cooperation in hunter-gatherers. Nature. 481:497–501.

Fowler JH. (2012). Behaviour: Life interwoven. Nature. 484(7395):448-449.

Gleiser PM, Spoormaker VI (2010). Modelling hierarchical structure in functional brain networks. Philosophical Transactions of the Royal Society A: Mathematical, Physical and Engineering Sciences. 368(1933):5633-5644.

Mitreva M (2012). Structure, function and diversity of the healthy human microbiome. Nature. 486:207-214.

Turnbaugh PJ, Ley RE, Hamady M, Fraser-Liggett CM, Knight R et al. (2007). The human microbiome project. Nature. 449(7164):804-810.

PART ONE: TRACKING THE PATH TO COMPLEXITY

"Duct tape is like the force. It has a light side, a dark side, and it holds the universe together."

— Oprah Winfrey

A brief history of the universe

In this section I will repeat the word relationship (or its derivatives and synonyms) many times in order to emphasize its importance in the formation of a world of increasing complexity.

Everything animate and inanimate in our universe is built up mainly of certain invisible grains called relationships. This is not a rhetorical, motivational, or metaphorical statement. It's real. All that there is and that we can (and perhaps cannot) perceive, is built upon relationships. Moreover, as we said earlier about complex systems, much of the universe is also built upon levels.

In our universe, two trends that apparently point in opposite directions exist. One of them point towards the direction of destruction; the other, towards the building of complexity. The first is represented by a physical quantity called entropy. The second, by three forces that govern matter at all levels. These three forces used to be four, and even today are often referred to as the strong force, the gravitational force, the weak force and the electromagnetic force. However, weak and electromagnetic forces have been unified by physicists in the electroweak model (Ribarik and Sustersic, 1985). Entropy can be fast and explosive in the process of disorganization, while the three forces construct and organize matter slowly, but patient and tenaciously. And while both trends are necessary for the maintenance of life, I will refer at first to these three forces which build structures that range from mere atoms and molecules to the most complex social systems.

Before the time in which our universe originated, there was emptiness. The type of emptiness to which I here refer is not that what we see in our daily lives. We will oftentimes say that "our governments are good for nothing", or that "there is nothing in the space that separates stars from one another". But while looking carefully at these "nothings," we find there is always something in them. Governments make something good every once in a while, and there is certainly this invisible fabric from which space and time are made of, even in the case of interstellar empty space. But in the 'nothing' of the primordial emptiness there was no space, no time, and really, no Nothing.

The moment of creation is known among cosmologists as the "Big Bang". There has been much controversy regarding this singular event. We can track that infinitesimal moment approximately 13,700,000,000 years ago, when all matter and energy were condensed in a dimensionless point of infinite density. The Three Forces were then not three but one: the so-called Superforce. Most events that determine how our universe is today, occurred during the first single second of existence. The point of infinite density began to expand as an explosion, accelerating its expansion (the so-called "Inflationary Universe Theory" (Goncharov et al., 1987; Gold and Albrecht, 2003). With its expansion, the huge initial temperature began to drop very rapidly.

Before the end of the first second of existence, energy in the early universe began to condense due to rapid cooling, forming a plasma or soup composed of a variety of elementary particles that were grouped into two basic types: fermions and bosons. Soon, fermions called "quarks" were related in groups of three to form larger particles: positively charged protons, and electrically neutral neutrons (Kurki-Suonio et al., 1990).

Soon after the first second, the Superforce was divided into the three fundamental forces: the Strong force, the Gravitational force, and the Electroweak force. Each one meant to rule different spatial scales. Just about three minutes after the universe came into existence, the strong force, which acts in a tiny range of distances, began to relate protons with each other and to neutrons to form simple atomic nuclei in a process called "nuclear fusion". Binding a proton to a neutron formed deuterium nuclei, while the binding of two protons and two neutrons originated atomic nuclei of helium. Many protons (most of them) were left alone, without any bonds, to form the nuclei of hydrogen, the simplest of all elements (Fields, 2006).

About 23 minutes after creation, the early universe ceased forming atomic nuclei due to cooling by adiabatic expansion. Most of the material was composed of hydrogen nuclei, followed by a considerable amount of helium and traces of heavier elements. The temperature was still too high to allow the formation of atoms. For the formation of these, positive nuclei (comprising protons) had to form bonds with negative fermions called electrons. These relationships were possible due to the action of the electroweak force (to be more precise, the electromagnetic component of the weak force).

Three hundred and seventy seven thousand (377,000) years since creation had passed before the electroweak force started to build atoms. During that time, the type of bosons which we call photons (that make up light and other electromagnetic radiation) could not move freely in the dense universe. Those were times of complete darkness, but around our universe's birthday number 377,000 the temperature had dropped enough to allow nuclei to trap electrons. The confinement of electrons to their places around the nucleus, thereby forming electrically neutral atoms, made the universe transparent allowing photons (or particles of light) to travel freely and unhindered in space (decoupling) (Hinshaw et al., 2009; Wall, 2012). And light appeared! However, it should be clarified that this first light was not coming from any particular point, but rather filled every space of this universe, expanding along with it. It was the flash of the initial instants of the universe. In 1978 Arno Allan Penzias and Robert Woodrow Wilson won the Nobel Prize in Physics for the discovery of the cosmic microwave background radiation, a 3 degree Kelvin radiation that fills every corner of the universe; the signature of the first glimpse: the big bang (Penzias and Wilson, 1965).

By the time the universe became transparent there were no stars nor galaxies yet, so there were no punctual sources of light. It was still very dark. A considerably long time had to pass before the work of our third force (gravity), the weakest of the three, capable of acting through enormous distances, became tangible.

The force of gravity began to approach and relate the different atoms to form clusters of matter. Gravity (gravitational force) would be responsible for progressively increasing the density of said gas clouds, approaching atomic nuclei to each other, and thereby increasing the temperature. When the nuclei were close enough to interact, the strong force could get back in action. The Strong force began to link these atomic nuclei thus activating nuclear fusion (Busso, 2010). About two hundred million (200,000,000) years after the origin of our universe, with the energy released by nuclear fusion, the dense condensations of matter lit up forming the first generation of stars (Bromm et al., 2009). These first stars (and the universe in general) were rich in hydrogen and helium. Gradually, as the stars burned due to nuclear fusion driven by the strong force, in their interior the atomic nuclei were bonding together, hence creating progressively heavier elements like carbon, nitrogen and oxygen—namely star nucleosynthesis— which would be essential for the subsequent formation of living organisms.

But gravity, additional to bringing the atoms together to give rise to the stars, also acted to a much larger scale, approaching large groups of stars to each other, originating the first galaxies. After a few billion years the first generation of stars was dying. Many, having exhausted their hydrogen and helium fuel, were cooling and finally exploding as supernovas, scattering their contents and thus enriching the interstellar space with the already formed elements (Andouze, 1995). The whole process was repeated. Again, the force of gravity would gather interstellar gas produced by the explosions of the stars of the first generation to form new clouds, then new stars, many of which now counted with planetary systems orbiting them thanks to the diversity of elements created in the first generation. About 4.6 billion

(4,600,000,000) years ago, halfway between the edge and the center of a galaxy which we now call the Milky Way, one of such clouds began to condense to give rise to our solar system (Bouvier and Wadhwa, 2010).

Baby Earth

Initially, in the heart of this fledgling solar system several condensations formed, but only one of them would be big and dense enough to be activated by nuclear fusion. The Sun was born. Other condensations formed the larger outer planets: Jupiter, Saturn, Uranus and Neptune. The remainder of the gas formed smaller bodies that just like the large planets, revolved around the Sun. Gravity kept all these bodies orbiting randomly around the central star, thus making them collide with each other, giving rise to the inner planets: Mercury, Venus, Earth and Mars (Committee on Grand Research Questions in the Solid-Earth Sciences, National Research Council, 2008).

Then something remarkable happened within the planets. The temperature of the star was extremely high, and this condition disabled the electromagnetic force from relating different atoms to form molecules (however, evidence points towards possible synthesis of organic molecules in some stars; Kwork, 2004). But in the relatively low temperatures of the planets individual atoms could interact with each other to form compounds of various kinds. The electromagnetic force, one of the components of the electroweak force, would then take control of evolution towards complexity. Thereafter, this force would be the driver of life.

The electromagnetic nature of relationships between atoms (called chemical bonds) would be of two basic types. In the first one, an atom has an electron that can be given away easily. The atom finds another atom that in turn likes to take electrons, and a perfect relationship arises. The atom that gives an electron is now positively charged, and the one that accepts the electron ends up with a negative charge. Both atoms remain together because of their opposite charges. From this type of coupling, ionic compounds emerged, such as salts, acids and bases. The best known of such compounds, as well as the most ubiquitous and essential for life on Earth is sodium chloride, or table salt.

The second type of relationship is more strong and stable. In this one the atoms, instead of giving their electrons, share them with other atoms in what is called a covalent bond, much stronger than the ionic one. Among all the elements formed in the first generation of stars, one proved to be a true wonder of the "relations" in the atomic world: Carbon. An atom of this element is able to share four of its electrons with other atoms. One of the simplest compounds formed by carbon is methane gas, which shares its four electrons with four hydrogen atoms. But a carbon atom can form covalent bonds with another carbon, and this one with a third one and so on, which facilitates the construction of large carbon chains. The diversity of types of chains made up of carbon atoms is potentially infinite. It is this enormous capacity of carbon atoms to relate to others, forming complex molecules, what proved crucial in creating that special new phenomenon called life (Orgel, 1998).

But even before the formation of the first carbon compounds in our Earth, a substance with interesting chemical properties appeared. The relation between its atoms are halfway between the ionic and covalent bonds, and their spatial arrangement allows the molecule to acquire an electric polarity, meaning that positive and negative charges are distributed differentially throughout the molecule. I'm talking about water, or H_2O. The electric dipole arrangement allows the negative part of the molecule to be attracted to the positive end of another. Then, two water molecules are linked together by a much weaker bond than ionic, one that chemists have called "hydrogen bond". The liquid water, through hydrogen bonds, forms a mesh or network of relationships that gives it most of its properties, such as a high boiling point, the ability to absorb lots of heat without substantially increasing its temperature, and its ability to dissolve substances, among others. Most water and carbon on earth might have had their origins in objects from the asteroid belt region (Committee on Grand Research Questions in the Solid-Earth Sciences, National Research Council, 2008; Paesani and Voth, 2009).

After its birth Earth was still very hot, although its temperature was much lower than that of the sun. Its surface was composed of an ocean of molten material where there was no liquid water. The gradual cooling determined the formation of a crust of volcanic rock, followed by the appearance of water vapor about 4.4 billion (4,400,000,000) years ago, when the newly born Earth had only around one hundred million (100,000,000) years of existence (Wilde et al., 2001). The huge amount of water vapor was mixed with carbon dioxide released from the volcanic rocks causing the most severe storm in the history of the planet. Water falling down formed a huge and single body. It rained for many millions of years, and in this way, about four billion (4,000,000,000) years ago a world where 90% of the surface was covered by liquid water appeared. It was one vast, single ocean (Valley et al., 2002).

Raise of life

Five hundred million (500,000,000) years after the emergence of the aquatic world, an intense volcanic activity on the surface of the planet separated the waters from the waters, thus, forming the continents and oceans. It is likely that life arose in the depths of this early ocean, and then from there, spread towards the surface. But complex carbon compounds must have appeared before life could develop.

Carbon molecules interacted with each other and with

nitrogen and hydrogen atoms in a variety of configurations forming the compounds necessary for life, such as nucleotides and amino acids. The prebiotic oceans must have been plagued by a variety of such molecules that would gradually increase in complexity. The amino acids linked together to form proteins, and nucleotides to form long RNA and DNA molecules. Over time, the association of proteins with RNA chains must have originated large self-replicating molecules, capable of constructing copies of themselves. This last feature would be essential for life. At present time, there is almost no remnant of that primeval molecular world (nearly all modern theories regarding the origins of life are still based on Alexander Oparin's ideas. See Oparin, 1952).

From this story we can see clearly that relations and cooperation are not inventions of humankind, but phenomena inherent to life, and furthermore, to the universe itself. Various structures, initially solitary, with the passage of time will gradually form links with their peers, slowly and imperceptibly, all guided by those three forces of nature.

Through this process, specialized proteins of increasing complexity were formed. These sophisticated molecular machines, within which molecules assemble and work in an orchestrated manner, would continue to specialize until the first cells arouse. Currently the Cell is considered the basic unit of life in our planet.

Researchers mark the beginning of life with the appearance of the first simple cells some 3.5 billion (3,500,000,000) years ago; therefore, it is said that before that time the world consisting of self-replicating molecules was prebiotic (before life). The sequence of events that led to those first cells is currently unknown, although we can say that cellular life began with the emergence of an envelope of fat and protein, called cell membrane, which protected the nucleus and cytoplasmic components from external environment (Peretó et al., 2004). The emergence of this barrier gave those first single celled organisms called prokaryotes (of bacterial nature) the ability to regulate its internal environment. The blossoming of bacterial life in the oceans must have been spectacular and of fierce competition. The only way of feeding was predation on other bacteria and on self-replicating molecules. It is probable that not only energy and nutrients were obtained from food, but genetic information as well in what is called Horizontal Gene Transfer (HGT). The molecular world was almost completely absorbed by the cellular world. Most of the phylogenetic information from bacteria, archea and protist/animals before and after the split into separate kingdoms must have been deleted by HGT. That makes the precise — or even approximate — reconstruction of the evolution of the cellular world very difficult, and theories to explain it remain highly speculative (for a review on HGT and the lost of genetic information read Woese CR, 2002).

Energy for replication and other processes of life in the presence of a reducing atmosphere, devoid of oxygen and rich in carbon dioxide, was produced exclusively by fermentation of organic molecules incorporated from the outside. Afterward, some cells developed the ability to use sunlight to make their own food. These photosynthetic bacteria (cyanobacteria) began to use carbon dioxide from the atmosphere, and as a result, oxygen began to be released some 2.8 billion (2,800,000,000) years ago.

One plus one is more than two

About 1.5 billion (1,500,000,000) years ago, a series of surprising relationships took place. Certain types of heat resistant bacteria (sulfidogenic archaebacteria) used to feed from swimming eubacteria, but over time, the two types would merge by means of symbiogenesis forming a single organism. This union originated a cell called archeaprotist, a kind of rudimentary amitochondriate eukaryote that possessed a well-defined nucleus.

These first eukaryotes were sensitive to oxygen and exposure to this gas destroyed them. The gradual increase in atmospheric oxygen concentration produced by photosynthetic bacteria led to a new relationship. Cells incorporated in its interior, through a second symbiotic relationship, small bacteria that had discovered a new technology: the utilization of oxygen for metabolism. The relationship between ancient anaerobic mobile cells and breathing eubacteria (α-proteobacteria) prove to be one of the most successful of all times. Modern mitochondria of eukaryotic organisms (including us humans) are the descendents of those eubacteria. Currently, they are the power plants of all our cells, for which they oxidize food to form carbon dioxide and water. This relationship allowed the new type of eukaryote to use atmospheric oxygen, giving it an evolutionary advantage (Gray et al., 1999).

In another association non-photosynthetic protists engulfed and enslaved a photosynthetic cyanobacterium, which came to be the organelles called plastids. This partnership gave rise to the eukaryotic cell ancestor of all green plants and algae (Bhattacharya et al., 2004). The idea of the emergence of modern eukaryotic cells by a series of symbiotic relationships between different types of bacteria was first proposed by Lynn Margulis and is called Serial Endosymbiotic Theory, or SET (Sagan, 1967; Margulis, 2004).

The emergence of the eukaryotic cell allows us to appreciate to the full extent the power of relationships. The ability to survive was increased exponentially in eukaryotes through the synergy of the components. Unicellular eukaryotes acquired many forms and dominated the earth for hundreds of millions of years. The intense photosynthetic activity during two billion (2,000,000,000) years mediated first by photosynthetic cyanobacteria and then by eukaryotic unicellular plants, increased gradually the atmospheric levels of oxygen, slowly turning our planet into a beautiful blue sphere (De Marais, 2000).

The abundant amounts of oxygen allowed proliferation of heterotrophic eukaryotes, i.e. those that lack chlorophyll and are unable to make their own food so they need to hunt and feed on other living things, usually using oxygen to obtain energy from them. Those would be the ancestors of multicellular organisms such as fungi and animals (including us). However, at present there are direct unicellular descendants of those beings, very similar to them, forming part of the kingdom Protista, the most diverse of the life kingdoms harboring heterotrophic and autotrophic species. The heterotrophic oxygen consumers are characterized by possessing a much more active metabolism and mobility than those of plant life. The development of structures for locomotion in protists quickly permitted them to venture and populate every corner of the planet (Taylor, 1980).

In a world of fierce and unmerciful competition the practical utility of relationships is evident. It is necessary to form partnerships; to learn to use resources elaborated by others, and as an exchange share with others the product of the own abilities. For an individual, it is convenient to live in a communal environment surrounded by peers who, to a greater or lesser extent, will supply many of the shortcomings. That is mainly why soon after the emergence of life, single-celled organisms acquired the ability to communicate and interact with their peers. The gregarious cells that tended to live together might have been favored by the evolutionary mechanism due to various factors including the protection offered by the group against predators, and the division of work.

Gradually, unicellular organisms were becoming more adapted to life in the community. Some eukaryotes began to form closer relationships joining together in small colonies or filaments. In some communities, gregarious life evolved to the point it became impossible, or at least very difficult, to survive isolated from the group. The specialization of individual organisms in specific tasks within the community gradually increased interdependence, with the subsequent increase of relationships, both in number and in quality. There was no turning back. The community of organisms was evolving to the next level of complexity: the multicellular organism.

At this point one can realize that there are no clearly defined boundaries between the unanimated and the biological worlds, as well as between unicellular and multicellular organisms. The transitions are determined by the continuous increase in complexity guided by the three forces acting in conjunction, as if they had a strange and mysterious agreement to increase the number of relationships and create increasing levels of structural organization. A sequence begins to emerge. An important technological innovation in communications or aggregation triggers a phase transition—spreads rapidly throughout the system—allowing better relations between elements, which in turn permit an explosion of life and new structures, more complex than before. From there evolution, while "choosing" the most adapted structures, experiences with many different variations until the next technological breakthrough—the one that will allow life to take the next big leap—appears, in a never ending cycle.

Chemical Communication

Some types of bacteria found an interesting and innovative way of communication between each other that allowed them to act collectively. These small creatures produce substances, called autoinducers, which can be detected by other bacteria. The bacterium that detects the substance then performs what has been called "quorum sensing," by which estimates the amount of bacteria of its own and of other species. This "knowledge" helps the organism to make "decisions" in agreement with its sisters (Miller and Bassler, 2001).

This type of communication can be seen in a variety of modern bacterial species which present different behaviors in isolation and when part of a group. Because the concentration of autoinducer increases according to the number of bacteria, this method allows them to "count" the number of sisters. Some behaviors occur only after autoinducer concentration surpasses a threshold. Upon reaching this limit entirely new synchronic group behavior arises. Some examples of synchronous behavior are bioluminescence, secretion of virulence factors, the formation of biofilms and the production of certain pigments (Antunes et al., 2010; Dickschat, 2010; Majumdar et al., 2012; Hornung et al., 2013).

This type of simple chemical communication drives the bacterial community into a group, coordinated behavior, acting as a whole, as an organism in its own right. Communication between bacteria gives us clues on how cells began to form multicellular organisms. However, bacteria never found the way, and the big jump was given by eukaryotes.

Evolutionarily speaking, when did the community of unicellular organisms became a single multicellular one? I must insist that a line is difficult (if not impossible) to define, but the transition must have been allowed by two innovations: in the first place, genetic variants that made cells prone to form cooperative cellular aggregates (Ratcliff, 2012). In second place, chemical intercellular communication through a distance in eukaryotes might have allowed structural organization through concentration gradients. Insights of this organizational mechanism can be found in the developing embryo as well as in bacterial biofilm formation by the aforementioned quorum sensing (Nusslein-Volhard, 1996; Gurdon and Bourillot, 2001; Fux et al., 2005).

The most popular type of intercellular communication resulted to be a "messenger" molecule called ligand, produced by a cell. The ligand travels, or rather diffuses through a space, and finally binds to a molecule (receptor) which is in the surface or inside of another cell. The development of a wide variety of chemical ligands allowed fluid communication between different cells. In the same way other successful genetic traits disperse in populations or useful inventions are rapidly adopted by human groups persisting in time, chemical communication might have been adopted promptly by unicellular organisms. Furthermore, it continued to be used throughout the entire evolution, from autoinducers to hormones, neurotransmitters, and growth and differentiation factors in more complex beings such as cockroaches or humans, and in general in the whole spectrum of life (as an example, review the evolution of the steroid receptors in humans in: Eick and Thornton, 2011). Chemical communication was one of the factors that allowed the leap from unicellular to multicellular life forms, but also continued to be used among multicellular organisms in order to communicate with each other and with their environment. Some species of social insects such as ants, exhibit complex group behavior through chemical communication. We humans use chemical communication with our environment or other humans to detect food, harmful substances, and even the perfect sexual mate, all by smell (chemical communication between humans or ants is carried out through ligands called pheromones. For example, Jacob et al., 2002 address the interesting relation between genetic variants of HLA in humans and woman's choice for male odor).

But chemical communication has a limitation: The speed in which a molecule diffuses from one cell to another is not very fast, and its range of action, relatively short. Therefore, a multicellular organism only availing this type of communication between its cells has a growth limit. The low speed at which signals travel between different regions of the body will render individuals above a certain size to respond slowly to ever changing ambient conditions. Size then imposes a restriction upon chemical communication on the evolution toward complexity of a multicellular organism.

But then again, the diversity of life found a new route, a new transportation technology. In some organisms certain cells gradually specialized and transformed to form tubes or channels which allowed faster transportation of chemical messengers between distant cells, as well as nutrients and waste interchange between inner cells and the external environment. Vascular systems allowed a progressive increase in the size of multicellular organisms (Wilkens, 1999). But this growth in size and complexity could not advance substantially, until the next innovation appeared.

Electric transmission

Some multicellular organisms began to experience interesting changes. Certain cells gradually specialized in the conduction of electrical impulses traveling along membranes by means of voltage-dependent ionic channels (Liebeskind et al., 2011; Widmark et al., 2011; Jensen, 2012; Ueya et al., 2012). With time, they grew longer and came into contact with each other in order to transmit signals between distant regions of the multicellular organism. The electric transmission between distant cells allowed for a faster response. The specialized cells began to form communication networks, originating the first nervous systems. Today, we can see these primitive nervous systems in form of lattices in coelenterates, multicellular radial organisms among which are the anemones and jellyfish. They are simple, and some of them transitional between the community of unicellular organism and the true metazoan (the hydrozoan or hydra is an example) (Petersen, 1990; Syed and Scherwater, 1997).

Metazoans appeared in our Earth some 540 million (540,000,000) years ago, during the so-called Cambrian explosion. The most primitive metazoans found in the fossil record correspond to the beginning of the Cambrian period. Among the probable causes of this explosion of life are: an increment in atmospheric oxygen accelerated due to the proliferation of plant life, the emergence of a group of developmental genes (called HOX), severe climate changes, strong competition for ecological niches, and the emergence of the protein collagen (Stanley, 1973; Hsu et al., 1985; Tucker, 1992; Knoll and Carroll, 1999). Indeed, innovations in distant intercellular communication must have been a crucial factor.

In some species, fuzzy neural networks as seen in coelenterates progressively transformed in more centralized condensations capable of processing information to some extent. Helminthes (worms) are a good example. Some species of flatworms have nervous systems composed of a primitive ringlike brain and cords joined by commissures. In general, the first central nervous systems consisting of cords first appeared in inferior chordates. This cord persisted along evolution in higher chordates, forming the spinal cords (Reuter and Gustafsson, 1995).

The emergence of dense neural centers allowed the formation of the circuitry for the basic processing of the information, but also led to the emergence of specialized cells in the detection of environmental conditions through stimulation by light, sound or gravity. The first sensory organs appeared. Some types of primitive eyes called *ocelli* did not even have the need for a center of information processing. Zooplankton larvae have light detecting cells that are directly connected to the natatorium apparatus of the animal, which simply follow the direction of the light source (Salvini-Plawen, 2008).

Gradually, neural cords in lower chordates suffered further condensations originating the first segmented nervous systems, which have nodes or ganglions that control the flow of information from each segment of the animal in a more complex way. Subsequently these structures evolved until the appearance of real brains, compartmentalized to manage the flow of information through several levels. It seems probable that brains emerge up to four independent times in evolution (Glenn-Northcutt, 2012).

The development of specialized structures for the detection of light, sound, gravity and environmental chemicals allowed the animal kingdom to develop a better response to the environment, and most important, increasingly conspicuous ways to communicate with their peers. The flowering of the variety of forms of communication between biological organisms cemented the bases for the next evolutionary leap, towards the level of communities of complex multicellular organisms.

It should be noted that the emergence of a new level of complexity does not imply the disappearance of prior levels. Each new level incorporates the others and thus, higher vertebrates show the entire evolutionary range in the same individual. Digestive systems of mammals, for example, have autonomous neural networks that resemble those in simple metazoarian. These networks serve as regulators of the digestive functions and maintain, to some extent, the digestive tract independent from the rest of the nervous system (Gershon, 1981). But mammals also have, in their nervous systems, simple condensations (nerves) and various processing centers with increasing complexity that correlate, caudal to rostral, to the time of their appearance in evolution, from the spinal cord, through the brainstem, to the cerebral cortex (Glenn-Northcutt, 2012).

Social insects and collective intelligence

I must say that this part of the essay is not intended to be an exhaustive review of the phylogeny of the nervous systems. Its main purpose is to emphasize the way in which progress in communications and relations between the elements of a population, whether of molecules, cells, multicellular individuals or communities of individuals, allows subsequent growth in size and complexity. In the animal kingdom, the most complex communities are those of social insects, many of which exhibit an intricate population structure with separation of tasks and complex social behaviors (complex behavior in ant societies is described in the work of EO Wilson, as for example: Wilson, 1963). Let's take the ants. Communication between the elements (ants) in the ant colony is mainly of four types: Chemical, tactile, visual and auditory. All functions in and out of the nest are finely regulated by means of these four types of signals.

Problem solving by means of coordinated behavior in ants or other social insects such as bees has been coined collective intelligence, i.e., an emergent behavior in a system composed of many individuals following a small number of rules. Collective intelligence makes the system behave as if it was a singular, inseparable unit. For a broader example we can observe the motor vehicle traffic in a city, which when seen from a certain altitude seems of an organized liveliness (Wolpert and Tumer, 1999).

I argue that the term "collective intelligence" is excessive for this type of "simpler" complex systems. They would be at best primitive, rudimentary and budding intelligences. When comparing an ant colony or perhaps an ancient human group with the phylogeny of multicellular organisms we realize that the same limitations of the first metazoans — such as coelenterates discussed earlier — apply to the nest, although in a higher evolutionary level. While very well structured, orderly behavior among individuals within the nest is dictated primarily through chemical signals, a slow method that requires proximity between participants. Visual, auditory and tactile communication also needs relative proximity.

Primitive and slow ways of communication may act as restriction factors limiting further evolution to complexity. Despite of the existence of giant ant nests, the maintenance of their structure is governed primarily by direct contact and chemical gradients. If we say that a jellyfish is intelligent, well, we could apply the term to the ant colony. Needless to say, the difference in intelligence between a primitive radial metazoan as a jellyfish and a higher mammal as an elephant, a dolphin or a human is abysmal. Similarly, to generate the leap from primitive collective intelligence of the ant community to the great communal intelligence it is necessary to develop faster communication methods that could act at greater ranges of distances. For the emergence of a true higher collective intelligence more complexity is needed; one of a type that does not exists in Earth yet, but which will probably find the way through the human species.

Memes and spoken language

Our own species is closely related to the great apes: chimpanzees, bonobos, gorillas, and orangutans. But unlike them, humans developed a vocal apparatus that allowed performing a wide range of sound combinations which conferred more fluid communication capabilities among members of a population. Whilst a variety of songbirds and marine mammals are able to articulate complex sounds with possible cultural connotations (Comins and Gentner, 2013; Cantor and Whitehead, 2013), this is a necessary but not sufficient characteristic for the development of a true complex language. In the other hand chimpanzees are holders of the abstraction level necessary for understanding language, especially gestural. However, they lack the structures required for the expression of more complex combinations of sounds, although it is quite possible that comprehension of linguistic sounds might have been already present in common ancestor of chimpanzees and humans (Taglialatela, 2011).

But we humans have both qualities fully developed: mental abstraction mechanisms, and the vocal apparatus. The development of the capacity of phonation is attributed to a variant of a gene called *FOXP2* (Fisher and Scharf, 2009; Ayub et al., 2013), which codes for a transcription factor that regulates the activation of around 100 other genes. *Homo neandertalensis*, a specie parallel to modern humans — and our ancestor as well, to some extend — that originated in Europe about 230,000 years ago had the same variant, for which reason researchers believe was able to communicate through speech (Kraus et al., 2007).The emergence of language required the development of two brain regions: Wernicke's area, involved in understanding and described as the seat of meaning and sound structure, and Broca's area, related to the emission of language and referred to as the seat of grammar (Fisher and Marcus, 2006). It is even possible that this structures of language had already been highly developed in the ancestor of humans, *Homo habilis*, more than one million years ago, as pointed by Tobias (1991) from studies on endocasts constructed from skulls which suggest the existence of a delimited Broca's area in that species.

Human intelligence along with our ability to understand and transform our environment is due to many concerting evolutionary factors acting altogether. Straight walking allowed a greater mobility of the upper limbs that in turn permitted the elaboration and manipulation of tools; stereoscopic and binocular vision aided in a more efficient performance in activities related to hunting; and decreased gestational time resulted in premature infants with a longer period of learning and higher brain volumes, a process called encephalization (Ciochon and Fleagle, 1987).

All of the above factors have been necessary in the emergence of humans as a successful species; however it is language that might have permitted the development of larger and more complex communities due to the cohesion that communication allows, transmitting abstract thoughts and feelings to peers in a simple way. With the ability to communicate ideas, hunting went bigger and more successful. It also permitted the planning of every step of the hunt game. Naming each of the varieties of fruits and roots might have facilitated the collection of food, and the new ability of expanded social engagement determined by language might have allowed communal child rearing by women, improving the survival of the young. Socialization through speech helped to extend the group's relations. On the other hand, the spoken word might have given impulse to acculturation.

All in itself, culture is a quality not restricted to humans. For example, the existence of a process of enculturation in chimpanzees has been demonstrated (Whiten, 2005). Groups of chimpanzees living in different regions acquire different skills by learning. However, with the development of spoken language acculturation received a huge boost by permitting the categorization of elements (nouns) and actions (verbs) in the everyday world, allowing conveying ideas more easily. In contrast, culture in chimpanzees is demonstrative. With the emergence of language life experienced a new big jump. Like baton in a relay race the task of storing and transmitting information from one generation to the next passed from genes into its equivalent in a higher level, namely memes (Dawkins, 2006), emergent elements of the complexity of the human brains. And as a result a huge part of evolution of living matter was transferred to new kinds of cultural evolution, such as technology.

Human genes continued to evolve and adapt, but memes were primarily responsible for subsequent human evolution towards complexity.

Metropolis and the written words

History was born with writing and a new era of human kind began. Writing enabled us to develop and maintain social regulating laws, instructions, and in general, allowed all the elements of culture to be accumulated and accurately transmitted from generation to generation. No wonder that writing was developed almost alongside with agriculture and livestock, which permitted a rapid expansion of the population. The first great civilizations flourished near rivers that were used as sources of water for irrigation, as Mesopotamia (now Iraq) between the Euphrates and the Tigris, the Egyptian civilization on the banks of the Nile, and the Chinese civilization along the Yellow river (Keightley, 1983; Daniels, 1996; Mitchell 1999).

Human populations increased in size thanks to the availability of food, but also to advances in transportation and communications. The use of pack animals provided transport and communication between individuals, who then on could communicate well through documents carried by messengers on horseback o by camel: the first curriers. Animals became an important part of trading relations between nations. Donkeys were amply used as pack animals in Egypt since the fourth millennium B.C. and in Ethiopia since 2,270 B.C. Historical records reveal the employment of camels with military purposes since the first millennium B.C. (Knauf, 1983; Blench, 2000). In Medieval Europe, riding emissaries replaced foot messengers allowing better and faster communication not only within the same city but between cities (Small, 1990). Trade and migration were activated by the advent of shipping and pack animals which allowed the transformation of city-states into great empires. The world began to shorten.

Navigation increased the range of human action. People could populate and conquer continents and the most remote places around the planet. Bamboo rafts were (probably) used in the peopling process of Southeast Pacific islands some 50,000 years ago (Horridge, 1995). Terrestrial or maritime migrations from Asia crossed the Bering Strait more than 15,000 years ago, populating North, Central and South America (Reich et al., 2012). Maritime transportation allowed sixteenth century Europeans to discover and conquer the Americas in what has been one of the biggest cultural shocks in recorded history (as for example, details on the extermination of Indo-American people in Honduras can be found in Newson, 1986). However, even though the world was more connected, populations still lived in relative isolation. The industrial revolution of the eighteenth and nineteenth centuries would lead to a series of inventions in the fields of communication that would pave the way to the next big leap: the global community.

The industrial revolution was the discovery of what nature knew since the dawn of life. Briefly, that specialization of individuals in a small part of the process is more efficient than everyone doing everything (Moore, 1959). Then the production line was born and with it, the economy of scales. The amount of goods produced increased and prices dropped which lead to a process of convergence of small autonomous units in a larger and structured one. Humanity began to experience an unprecedented rural to urban exodus, a phenomenon known as urbanization, causing many cities worldwide to surpass the threshold of 10 million inhabitants during the 20th Century (Kabisch and Haase, 2011; McCann and Acs, 2011).

In the industrialized world (and in the developing world afterwards), rapid transportation allowed for more efficient distribution of goods and the transfer of workers from their homes to places of production thanks to the invention of the steam engine and the internal combustion engine, boosting urbanization and development (Abrams and Mulligan, 2009; Jordan, 2011; Ogun, 2010). The cities gradually became complex centers of production characterized by networks along which the high-speed oil driven machines transited. Complexity was rapidly accumulating to the community level, namely, the big city. Metropolis is the new human superorganism.

Meanwhile, these superorganisms have approached each other due to the transport of goods and people at speeds hard to imagine only a few generations ago. Finally, the golden dream was fulfilled and humankind venture into the air. Airborne transportation joining cities is one of the factors leading to the global superorganism, and its offspring, interplanetary flights, may allow the future settlement of other worlds, and thus, the development of communities of global (planetary) superorganisms in an interplanetary transport network. When Wilbur and Orville Wright launched themselves as eagles to the heights on the plains of Kitty Hawk never imagined they were preparing the ground to cut the umbilical cord that connects us with our planet. In turn, we must remember that our path towards complexity started within the stars, with the creation of the necessary elements for life. The transition to more complex levels shall take us back to them.

LITERATURE CITED

Abrams BA, Li J, Mulligan JG (2009). The Steam Engine and US Urban Growth During the Late Nineteenth Century. *Working papers series.* University of Delaware. (No. 09-06).

Andouze J, Silk J (1995). The first generation of stars: first steps toward chemical evolution of galaxies. *The Astrophysical Journal Letters.* 451(2):L49.

Antunes C, Ferreira R, Michelle, Buckner M, Finlay B (2010). Quorum sensing in bacterial virulence. *Microbiology.* 156(8): 2271-2282.

Ayub Q, Yngvadottir B, Chen Y, Xue Y, Hu M et al (2013). FOXP2 targets show evidence of positive selection in european populations. *The American Journal of Human Genetics*. 92(5):696–706.

Bhattacharya D, Yoon HS, Hackett JD (2004). Photosynthetic eukaryotes unite: endosymbiosis connects the dots. *Bioessays*. 26(1):50-60.

Blench RM (2000). *A history of donkeys, wild asses and mules in Africa.* In: *The origins and development of African livestock: Archaeology, genetics, linguistics and ethnography,* 339-354. UCL Press.

Bouvier A, Wadhwa M (2010). The age of the solar system redefined by the oldest Pb-Pb age of a meteoritic inclusion. *Nature Geoscience*. 3:637–641.

Bromm V, Yoshida N, Hernquist L, McKee CF (2009). The formation of the first stars and galaxies. *Nature*. 459(7243):49-54.

Busso M, Maiorca E, Magrini L, Randich S, Palmerini S et al. (2010). News from Low Mass Star Nucleosynthesis and Mixing. *arXiv preprint* arXiv:1012.2546.

Cantor M, Whitehead H (2013). The interplay between social networks and culture: theoretically and among whales and dolphins. *Philosophical Transactions of the Royal Society B: Biological Sciences*. 368(1618): 20120340-20120340.

Ciochon RL, Fleagle JG (1987). *Primate evolution and human origins.* New York: Aldine de Gruyter.

Comins JA, Gentner TQ (2013). Perceptual categories enable pattern generalization in songbirds. *Cognition.* 128(2):113-118.

Committee on Grand Research Questions in the Solid-Earth Sciences, National Research Council(2008). *Origin and Evolution of Earth: Research Questions for a Changing Planet.* The National Academies Press. Washington D.C.

Daniels PT, Bright W (1996). *The world's writing systems (Vol. 198).* New York: Oxford University Press.

Dawkins R (2006). *The selfish gene* (No. 199). Oxford: Oxford University Press.

De Marais DJ (2000). Evolution. When did photosynthesis emerge on Earth? *Science.* 289(5485):1703-5.

Dickschat JS (2010). Quorum sensing and bacterial biofilms. *Natural Products Report.* 27: 343-369.

Eick GN, Thornton JW (2011). Evolution of steroid receptors from an estrogen-sensitive ancestral receptor. *Molecular and Cellular Endocrinology.* 334(1–2):31–38.

Fields BD, Olive KA (2006). Big bang nucleosynthesis. *Nuclear Physics A.* 777: 208-225.

Fisher SE, Marcus GF (2006). The eloquent ape: genes, brains and the evolution of language. *Nature Reviews Genetics.* 7(1):9-20.

Fisher SE, Scharff C (2009). FOXP2 as a molecular window into speech and language. *Trends in Genetics.* 25(4):166-177.

Fux CA, Costerton JW, Stewart PS, Stoodley P (2005). Survival strategies of infectious biofilms. *Trends in microbiology.* 13(1): 34-40.

Gershon MD (1981). The enteric nervous system. *Annual Review of Neuroscience.* 4:227-272.

Glenn-Northcutt R (2012). Evolution of centralized nervous systems: Two schools of evolutionary thought. *Proceedings of the National Academy of Sciences (USA).* 109(Supplement 1): 10626-10633.

Gold B, Albrecht A (2003). Next generation tests of cosmic inflation. *Physical Review D.* 68(10):103518.

Goncharov AS, Linde AD, Mukhanov VF (1987). The global structure of the inflationary universe. *International Journal of Modern Physics A.* 2(03):561-591.

Gray MW, Burger G, Lang BF (1999). Mitochondrial evolution. *Science.* 283(5407):1476-1481.

Gurdon JB, Bourillot PY (2001). Morphogen gradient interpretation. *Nature.* 413(6858):797-803.

Hinshaw G, Weiland JL, Hill RS, Odegard N, Larson D et al. (2009).

Five-year Wilkinson Microwave Anisotropy Probe observations: Data processing, sky maps, and basic results. *The Astrophysical Journal Supplement Series.* 180(2): 225.

Hornung C, Poehlein A, Haack FS, Schmidt M, Dierking K, et al. (2013). The Janthinobacterium sp. HH01 Genome Encodes a Homologue of the V. cholerae CqsA and L. pneumophila LqsA Autoinducer Synthases. *PLoS ONE.* 8(2):e55045.

Horridge A (1995). *The Austronesian Conquest of the Sea – Upwind.* In: *The Austronesians: historical and comparative perspectives,* 134-51.

Hsu KJ, Oberhänsli H, Gao JY, Shu S, Haihong C, Krähenbühl U (1985). 'Strangelove ocean' before the Cambrian explosion. *Nature* 316: 809-811.

Jacob S, McClintock MK, Zelano B, Ober C (2002). Paternally inherited HLA alleles are associated with women's choice of male odor. *Nature Genetics.* 30(2):175-9.

Jensen MO, Jogini V, Borhani DW, Leffler AE, Dror RO et al. (2012). Mechanism of voltage gating in potassium channels. *Science Signaling.* 336(6078):229.

Jordan AL (2011). The Historical Influence of Railroads on Urban Development and Future Economic Potential in San Luis Obispo. *Master's Thesis. California Polytechnic State University - San Luis Obispo.*

Kabisch N, Haase D (2011). Diversifying European agglomerations: evidence of urban population trends for the 21st century. *Population, space and place.* 17(3):236-253.

Keightley DN, Barnard N (1983). *The origins of Chinese civilization (Vol. 1).* Univ of California Press.

Knauf EA (1983). *Midianites and Ishmaelites.* Sawyer and Clines. 147-162.

Knoll AH, Carroll SB (1999). Early Animal Evolution: Emerging Views from Comparative Biology and Geology. *Science.* 284(5423):2129-2137.

Kurki-Suonio H, Matzner RA, Olive KA, Schramm DN (1990). Big bang nucleosynthesis and the quark-hadron transition. *The Astronomical Journal.* 353:406-410.

Kwork S (2004). The synthesis of organic and inorganic compounds in evolved stars. *Nature.* 430(7003):985-991.

Liebeskind BJ, Hillis DM, Zakon HH (2011). Evolution of sodium channels predates the origin of nervous systems in animals. *Proceedings of the National Academy of Sciences USA.* 108(22):9154-9159.

Majumdar S, Datta S, Roy S (2012). Mathematical Modeling of Quorum Sensing and Bioluminescence in Bacteria. *International Journal of Advances in Applied Sciences.* 1(3): 139-146.

McCann P, Acs ZJ (2011). Globalization: countries, cities and multinationals. *Regional Studies*. 45(1):17-32.

Miller MB, Bassler BL (2001). Quorum Sensing in Bacteria. *Annual Review of Microbiology*. 55:165-199.

Mitchell L (1999). Earliest Egyptian Glyphs. *Archaeology*. 52(2). Available in the internet at:

http://archive.archaeology.org/9903/newsbriefs/egypt.html. Retrieved December 2012.

Moore FT (1959). Economies of Scale: Some statistical Evidence. *Quarterly Journal of Economics* 73(2):232–245.

Newson LA (1986). *The cost of conquest: Indian decline in Honduras under Spanish rule (Vol. 20)*. Boulder, Colorado: Westview Press.

Nusslein-Volhard C (1996). Gradients that organize embryo development. *Scientific American* 275(2):54.

Ogun TP (2010). Infrastructure and poverty reduction: implications for urban development in Nigeria. *In Urban Forum. Springer Netherlands*. 21(3):249-266.

Oparin AI (1952). *The Origin of Life*. New York: Dover.

Orgel LE (1998). The origin of life—a review of facts and speculations. *Trends in Biochemical Sciences*. 23(12):491–495.

Penzias AA, Wilson RW (1965). A Measurement of Excess Antenna Temperature at 4080 Mc/s. *Astrophysical Journal Letters.* 142:419–421.

Peretó J, López-García P, Moreira D (2004). Ancestral lipid biosynthesis and early membrane evolution. *Trends in Biochemical Sciences.* 29(9):469-477.

Petersen KW (1990). Evolution and taxonomy in capitate hydroids and medusae (Cnidaria: Hydrozoa). *Zoological Journal of the Linnean Society.* 100(2):101-231.

Ratcliff WC, Denison RF, Borrello M, Travisano M (2012). Experimental evolution of multicellularity. *Proceedings of the National Academy of Sciences USA.* 109(5):1595-1600.

Reich D, Patterson N, Campbell D, Tandon A, Mazieres S et al. (2012). Reconstructing native American population history. *Nature.* 488(7411):370-374.

Reuter M, Gustafsson MKS (1995). The flatworm nervous system: Pattern and phylogeny. In: The Nervous Systems of Invertebrates: An Evolutionary and Comparative Approach. *Experientia Supplementum.* 72: 25-59.

Ribaric M, Sustersic L (1985). A classical model of unified electroweak forces — I. *Il Nuovo Cimento A Series II.* 88(3):325-349.

Sagan L (1967). On the origin of mitosing cells. *Journal of Theoretical*

Biology. 14(3):225–274.

Salvini-Plawen L (2008). Photoreception and the polyphyletic evolution of photoreceptors (with Special reference to Mollusca). *American Malacological Bulletin.* 26(1-2): 83-100.

Small CM (1990). Messengers in the County of Artois, 1295-1329. *Canadian Journal of History,* 25(2):163-175.

Stanley, SM (1973). An ecological theory for the sudden origin of multicellular life in the late Precambrian. *Proceedings of the National Academy of Sciences (USA).* 70:1486-1489.

Syed T, Schierwater B (2002). Trichoplax adhaerens: discovered as a missing link, forgotten as a hydrozoan, re-discovered as a key to metazoan evolution. *Vie et Milieu.* 52(4):177-188.

Taglialatela JP, Russell JL, Schaeffer J A, Hopkins WD (2011). Chimpanzee vocal signaling points to a multimodal origin of human language. *PLoS One.* 6(4):e18852.

Taylor FJR (1980). On dinoflagellate evolution. *Biosystems.* 13(1–2):65-108.

Tobias PV (1991). *The skulls, endocasts, and teeth of Homo habilis. Vol. 4.* Cambridge University Press.

Tucker ME (1992). The Precambrian–Cambrian boundary: seawater chemistry, ocean circulation and nutrient supply in metazoan evolution, extinction and biomineralization. *Journal of the Geological Society.* 149(4):655-668.

Ueya N, Shirai O, Kushida Y, Tsujimura S, Kano K (2012). Transmission mechanism of the change in membrane potential by use of organic liquid membrane system. *Journal of Electroanalytical Chemistry. 673*:8-12.

Valley JW, Peck WH, King EM, Wilde SA (2002). A cool early Earth. *Geology.* 30(4):351-354.

Wall M (2012). Ancient galaxy may be most distant ever seen. Available in the internet at http://www.space.com. Retrieved: December, 2012.

Whiten A (2005). The second inheritance system of chimpanzees and humans. *Nature.* 437(7055):52-55.

Widmar J, Sundström G, Daza DO, Larhammar D (2011). Differential evolution of voltage-gated sodium channels in tetrapods and teleost fishes. *Molecular biology and evolution.* 28(1):859-871.

Wilde SA, Valley, JW, Peck WH, Graham CM (2001). Evidence from detrital zircons for the existence of continental crust and oceans on the Earth 4.4 Gyr ago. *Nature.* 409:175–178.

Wilkens JL (1999). Evolution of the Cardiovascular Systems in Crustacea. *American Zoologist.* 39 (2):199-214.

Wilson EO (2012). *The social conquest of earth.* Liveright.

Woese CR (2002). On the evolution of cells. *Proceedings of the National Academy of Sciences of the United States of America.* 99(13):8742–8747.

Wolpert DH, Tumer K (1999). An introduction to collective intelligence. *arXiv preprint* cs.LG/9908014.

PART TWO: THE ELEMENTS OF EVOLUTION TO COMPLEXITY

"Simplicity is the ultimate sophistication."

— Leonardo da Vinci

Electric transmission and availability of information

Let's keep in mind that communications between people in distant places before the industrial age were carried out by messengers on horseback, donkeys or camels. But a series of discoveries regarding electricity would open the door to the real global community. In 1832 Samuel Finley Breeze Morse, Joseph Henry and Alfred Veil invented a device that transmitted electrical signals between two stations, and by the end of the nineteenth century most of the world was interconnected by telegraph (Hochfelder, 2010). Electrical transmission was a real revolution, and soon, Alexander Graham Bell invented the telephone enabling the transmission of voice (although there is controversy regarding who invented it: Beauchamp, 2010).

The acknowledgment of the fact that broadcasting of images and sound to the people could be done without a physical medium through low-frequency electromagnetic waves gave birth to the mass media (radio and television). But the real communications revolution originated with the invention of the internet and the World Wide Web (Leiner et al.,2009; Hendler and Berners-Lee, 2010). The network allows for the universal access to knowledge, the latest news, and an increase in the number of relationships. The internet enables entirely new ways to unleash personal creativity. All that is required is a suitable device and network connectivity. Today these connections are becoming increasingly ubiquitous.

I oftentimes hear some criticism of the way in which we are bombarded with information. We have the television, radio, and now the Internet making it ever more difficult for us to decide what to see and listen, and what not. It is a fact that from the vast amount of data we have at our disposal, we only use a tiny percentage. So, why so much information if most of it will not be used? Is that not a waste of resources?

To illustrate that the total amount of information is not a waste of resources, but on the contrary, it should be available to all elements of a system for the subsequent evolution toward higher complexity I will refer to a similar case at a lower level. Multicellular organisms have all the information needed for their—or I should say our—own construction and operation stored inside their cells. The acquisition of genetic information has cost billions of years of trial and error and is fully compacted into the nucleus in the form of DNA. Now, a particular cell only needs a small part of that information to survive and exert its role in the organism. However, each of the cells of the multicellular organisms has the total DNA information available. This allows a cell to have the potential to perform every cellular function. Nevertheless, during the process of cell differentiation most of the genome is silenced by means of gene methylation or synthesis of small RNAs (Moazed, 2009; Raynal et al., 2012). Similarly, the fact that the universal information is available to every human lets us easily select the one that will be to our benefit. It is the accessibility to the universal information paired with human inventiveness, specialization and collaboration what is causing knowledge and progress to expand at an exponential rate.

Another notable example is language. A given population has at its disposal an extraordinary wealth of words; however, not all words are used by all people and this is most evident in technical languages. A specialist in the practice of his profession uses only the words that correspond to that particular area of knowledge and not others, although the full set of technical terms from all fields is available through books or the internet. We "silence" most of our language and keep what is needed in our daily social and work transactions. Thus, we see here how language reflects the structuring process given by specialization (division of labor) resulting in more complex societies, but the full and complete language is available to all.

Humans in modern society are slowly becoming like cells of a multicellular organism. All the information is available to every cell but each one uses only a fraction, and it is interconnected in some way with every other cell. As a cell depends on other cells for survival, we depend on our fellows for our own. In the evolutionary process of integration to the body the cell loses most of its functions to specialize in a few. As the global community grows, we are becoming less able to survive on our own. Like the cells that make up our bodies, we become increasingly interdependent (Vespignani, 2010). Each of us is only a small part of a superorganism.

Fractality

Fractals are irregular objects to which classical geometric rules cannot be applied, and additionally, exhibit self-similarity or power-law behavior. That is, if we take one of its parts it will be similar to the whole. Fractals best known are those of the Mandelbrot series (discovered by awarded Polish mathematician Benoît Mandelbrot), produced by algorithms using simple equations (Mandelbrot, 1982). Fractality is a common feature of many natural objects, and applications in biomedical sciences have been explored to some extent (Grizzi et al., 2007; Lopes and Betrouni, 2009; McNally and Mazza, 2010; Xing et al., 2012). Mandelbrot himself referred to the cauliflower as an example. If you cut off a piece of this vegetable you will see that it bears extreme similarity to the total. If you then cut back that part a bit, a new miniature cauliflower will appear, and this pattern will continue until some scale constraints begin to appear.

There are various types of fractality according to the extent to which self-similarity is respected. A fractal generated by a computer program from an equation can follow fixed rules and as a result will present exact self-similarity. In natural forms exactness is usually absent and so we have quasi-similarity (such as the cauliflower of Mandelbrot's example). In statistical fractality self-similarity is just a trend at different scales. Levels of complexity in life and its derivative systems exhibit quasi-similarity, statistical similarity, or (especially) multifractality, in which a single fractal dimension is not enough to explain the dynamics of the whole system (Spencer, 2009). Phenomena and behaviors observed at a level of complexity resemble in some manners those occurring at another level.

Informative elements and phylogenetic trees. Let's take a gene (call it gene A) from the human genome and sequence it completely, i.e., read the string of letters (that can be of four types, A, T, C and G) in the DNA fragment that contains the gene, as if it were a page from a book. Now, let's find a gene with a similar sequence in another species, such as a mouse. It results, not strangely, that gene A in the mouse has the same function as gene A in humans. Let's do the same with several other species, and then compare the sequences of gene A in all of them. From the resemblance of the sequences — i.e. the number of places that share the same letter — between pairs of species we can construct a phylogenetic tree that will reveal to us their relatedness, hence, the evolutionary history of those species. Two branches of the tree represent two species, and the trunk from which they stem from is the common ancestor. As we go down to the thicker, common branches, and finally the common trunk we are moving backwards in time. These genes with comparable functions and a high degree of likeness in their sequences (homology) among species are called orthologs (Frazer el al., 2003; Small et al., 2004).

Now let's take a human genome and sequence it thoroughly. Then, let's compare all the genes in that genome. It will turn out that many among them share varying degrees of homology. It is possible that gene A might be more similar to gene B than to gene C, so a higher degree of kinship between A and B is inferred. Based on that fact — homology among genes within a genome — we can build a phylogenetic tree since some genes have originated from others by duplication. That is, two specific genes may have a common ancestor gene that at some historical point duplicated, originating two identical copies. Over many generations both became increasingly differentiated from one another, and after many more generations, both might have evolved to occupy different functional niches. Genes in the same species that descend from a common ancestor are called paralogs (Fitch, 1970; Koonin, 2005; Gabaldón and Koonin, 2013).

It follows that the evolutionary process that we see between species is repeated when we look within a single species. Just as human paralog genes reflect the evolutionary history of the diversity of proteins within our body, the orthologs reflect the evolutionary history of the diversity of species. The process of evolution at the molecular level is a reflection of the process of evolution at the species level.

To put it in a more graphical manner and better appreciate the idea of fractality suppose we construct a phylogenetic tree of the many species that form an ecosystem using complete genomes. At the extreme of each branch we will have the complete genome of each species. But a single genome, in turn, is structured as a phylogenetic tree using paralogous genes within it. If we zoom in the branches we will see that each of those species resembles the tree of the entire ecosystem.

In an upper level of complexity, several cultural features share the same evolutionary progression experienced by genes. Surnames in most western countries are inherited by the father, so their frequencies can be used by geneticists to estimate the structure and dynamics of populations through techniques similar to those used for the study of Y chromosome markers. Some surnames have their origin from the modifications of others due to errors which resemble mutations. Additionally, when two populations come together the abundance of surnames increases, resembling the increment of heterocygocity in the case of genes. That is why the frequency of surnames can be used to compare similarities among populated units, such as villages, cities, and countries, which allow the construction of phylogenetic trees similar to those obtained from genetic data through mathematical methods based on isonymy (the probability of marriage between two persons with the same surname) (Crow and Mange, 1965; Jobling, 2001; Colantonio et al., 2003; Scapoli et al., 2007; Herrera Paz, 2013).

Vocables that compose a dialect change slightly from one generation to another. Languages usually expand themselves in each generation with recently invented expressions. Many words originate from new discoveries and inventions, but others begin their "lives" as vulgar and humble, within the hordes of young people or in the jargons, climbing the social ladder throughout generations until they become part of formal language just like the frequency of a mutation increases because of random forces or natural selection. Some others progressively decrease in usage until they finally disappear. Languages, like genes, are enriched with the fusion of two populations. Because phonemes and words evolve similarly to genes phylogenetic trees constructed from linguistic data tend to highly correlate with those constructed from DNA sequences (Cavalli-Sforza et al., 1988; Cavalli Sforza, 1997; Dediu, 2013). In addition to surnames and languages, there are many other examples of informative elements in cultural evolution that may resemble genetic, such as customs, tools, traditions and religion (Goodenough, 1997; Mesoudi et al., 2004; Mesoudi et al., 2006).

All major religions share similar moral and spiritual principles (Schwartz, 1995), but they all evolved to become different. Rituals, beliefs, or even exegesis (i.e. Interpretation of sacred texts) share common descent. For instance, Christianity and Judaism share a common trunk, but the former split in many forms such as Roman Catholicism, the Greek Orthodox Church, Presbyterianism, Congregationalism, Quakerism and Evangelism. Rituals and beliefs differ slightly even within each Church, reflecting recent divergence (Smith, 1990). Moreover, the fusion of cultures may result in the admixture of religious elements referred to as syncretism (Stewart, 1999).

Resemblance of the evolution of genes with informative elements at the cultural level may seem evident. This connection originates from the fact that populations tend to split and evolve separately; however, most phenomena at different levels of complexity show a more subtle type of similarity that may fit the multifractality model.

Selfish cells, loving cells. Selfishness is considered, from a moral standpoint, as one of the most negative human feelings to the extreme of being cataloged by the Catholic Church as a deadly sin. However, a dose of selfishness is necessary for survival, and selfish and altruistic behaviors must be in an approximate equilibrium imposed by the struggle between individual and group selection (Wilson and Wilson, 2007; Sober and Wilson, 2011). If that is the case, altruism can be seen as selfishness at a more abstract level, determined by group selection (Rachlin, 2002). But aside from altruism as a type of group selfishness, when the individual manifestation of this behavior (selfishness) surpasses a threshold the risk of destroying the body or society to which the selfish entity belongs is high, so, we might say that the system entered a state of disequilibrium with insufficient control mechanisms to return to normal operation.

I live in the city that is now considered the most violent in the world. My home city San Pedro Sula, as well as the rest of my country, used to be an oasis of peace. During the cold war in the nineteen eighties three Central American countries, El Salvador, Nicaragua and Guatemala, were plunged into bloody ideological revolutions while Honduras, located at the center of the isthmus, maintained an almost absolute tranquility. But the Cold War ended and with it the revolutions. During the nineties we in Honduras (as well as the rest of Central America) observed the emergence of small groups of juvenile offenders forming gangs which were then called "maras". They began to recruit other young men to a point the size of the groups was so big that it became impossible to control despite the implementation of strong anti-gang policies and bills, such as *"Mano Dura"* (Firm Hand) in El Salvador, "Zero Tolerance" and *"Antimara"* law in Honduras, and *"Plan Escoba"* (Operation Broomsweep) in Guatemala (Rodgers el al., 2009).

Moreover, drug trafficking and aggressive groups that migrated from other Central and South American countries began to spread to Honduras. Cocaine trafficking and money laundering were proved to be highly profitable enterprises, and the recruitment of young people spread epidemically. Law enforcement derived from the antidrug war in Mexico and Colombia displaced a substantial part of the drug trafficking operations to Central America, especially Honduras. Today, a significant proportion of the young adult population is involved in some drug-related activity, and many important local entrepreneurs are subsidized by drug traffic or involved in money laundering activities (Dudley, 2010). Violence has spread so much that it threatens to maintain the country in utter poverty. Recession was triggered, not because of the fear of spending money, but because of the fear of walking the streets.

It's hard to avoid comparing *maras'* and drug dealers' behavior with that of a self-destructive disease called cancer. Both—criminal activity and cancer—are based on elements present in unbridled selfishness; cells that refuse to restrain their impulse to divide uncontrollably in the case of cancer, and young criminals willing to enrich themselves at the cost of the lives of others, in the case of gangs. Like the gang, once the population of malignant cells surpasses a growth limit, cancer expands and treatment becomes extremely hard and habitually unsuccessful.

External threats as the case of attempted invasion of a country resemble bacterial infections. Fortunately, our bodies have flexible and alert immune systems ready to fight the latter. As countries have soldiers patrolling their borders, our skin and other tissues are patrolled by dendritic cells and macrophages which do not hesitate to activate the alarm upon the recognition of an invader (pathogen) by means of sophisticated detection mechanisms (Kumar et al., 2011). Promptly, the army responds. Macrophages and T and B lymphocytes lead the war. Soldiers (neutrophils) leave the comfort of their barracks and patrol areas to melee with the invader. As the soldiers of a country, these warriors of the flesh are willing to give up their lives for their peers, and as those, they count with a powerful arsenal.

The struggle between police corpses and crime is an ongoing competition. Gangs and drug traffickers constantly change their strategies to circumvent the authority. Similarly, microorganisms that make us sick mutate and hide to evade the immune system (Pierce and Miller, 2009; Lin and Shuai, 2010; Sorci et al., 2013). Sometimes, a group of immune cells attacks its own body, initiating an autoimmune disease (albeit several mechanisms for the emergence of these diseases have been proposed, a complete understanding of their genesis remains elusive; however, new RNA and genomic technologies are promising; Pascual et al., 2010). Regardless of the subjacent mechanisms in autoimmune diseases, similarities with corrupt police officers and government officials working for organized crime seem to me staggering.

Immune systems might have developed early in evolution. Bacteria and Archaea count with defense weaponry against bacteriophages—the viruses that infect and destroy them—such as restriction endonucleases, true molecular knifes that cut the invader's nucleic acids. More recently, a type of adaptive RNA-based immune system in bacteria has been discovered (Horvath and Barrangou, 2010; Stern and Sorek, 2011; Wiedenheft et al., 2012). It is assumed that in metazoarians immune systems evolved progressively from the use of general defense weaponry (innate immune systems) to sophisticated and customized gadgets for detection and destruction, such as T cell receptors and antibodies, capable of choosing specific targets (adaptive immune systems) (Cooper and Alder, 2006). The comparison of these systems with an army or a police force is more than a simple analogy. Both have the same functions on two different levels of complexity. Armies are social inventions, and immune systems, biological. The first use advanced electromagnetic communication devises to act as a whole, while the latter are mostly communicated through chemical ligands called cytokines (Pestka et al., 2004; Huising et al., 2006; Nomiyama et al., 2010). But their respective roles and the strategies they use are identical in principle.

On the other side of the spectrum, we encounter passionate love. Neural and endocrine mechanisms routed to mating are very powerful in nature. In humans, amatory behavior is driven by hormones and neurohormones such as testosterone, estrogen, and oxytocin (Pfaus et al., 2001; Tetel and Pfaff, 2010; Magon and Kalra, 2011). Generally, in most animal species the male actively seeks the female competing with other males for her favors, while it's the female who ultimately makes the choice (Hunt et al., 2009), but no matter who chases who finally it all ends up in copulation. This behavior is reflected at the cellular level where a huge contingent of active sperm is urged to fertilize an egg, but only one of them succeeds. Moreover, besides competition intricate interactions among spermatozoids such as sperm conjugation, a type of altruism, have been described (Higginson and Pitnick, 2011). The physical approach of males toward females and the underlying population interactions and strategies, resemble fertilization.

Sexual reproduction predominates in the living landscape of planet Earth despite its high costs in energy and time consumption. Apparently, natural selection is enhanced with genetic variability, and a population that reproduces sexually adapts faster than a non-sexual. Sex guarantees variability, especially in long genomes, and this is achieved by means of gene shuffling in the individual and the cellular levels, but also at the molecular level. Cellular divisions that give rise to the egg and the sperm are called meiosis. Within the cell nucleus, during meiosis, one chromosome inherited from the mother is paired with its counterpart, inherited from the father. Then, both come together, unite in an intimate embrace and exchange their material in a kind of mating referred to as genetic recombination which originates brand new hybrid (recombinant) chromosomes (Hadany and Comeron, 2008; Stower, 2012). In mechanisms that living organisms use to express erotic love and reproductive behavior, as well as in those used to defend themselves and maintain order — among many others — self-similarity becomes evident.

Students usually marvel at the similarity between Bohr-Rutherford's description of the atom structure and a planetary system. Even though it might be an oversimplification (Jeknic-Dugic et al., 2012), behind the existence of both type of systems we found identical fundamental principles: a central mass holding smaller masses by means of natural forces, which in turn might be different expressions of a single superforce operating at different scales of complexity. For my part, I marvel at the way in which a human brain can serve as a mirror — or perhaps as a sounding board — of the entire universe. That can only be achieved if the complexity of the human brain itself equals to that of the entire universe, but I suspect the comparison is more than a simple analogy.

Separation, differentiation and reunion

The construction of phylogenetic trees from informative

elements in a population is possible because of differentiation with the subsequent genesis of variability, which in turn is necessary for evolution, from the pure Darwinian point of view as well as for increment in complexity. Recall the formation of the first eukaryotic cells, which then evolved to the present multicellular organisms. The series of events were as follows: Bacteria expanded its population reaching all corners of the Earth. Random genetic mutations coupled with other forces such as natural selection induced differentiation and led to a large number of species, each with special abilities and adaptive features. When three of these bacterial species joined in symbiotic relationships the eukaryotic cell originated, something completely different, with totally new features or emergent properties. It happens here that the triad "isolation (or separation, if you prefer) → differentiation → reunion" is in many cases the key to evolution towards complexity, and we observe this continuously through the phylogenetic history of many organisms.

One of the prevailing theories regarding peopling of the world by *Homo sapiens,* the so-called "Out of Africa," tells us that early humans evolved in Africa and from there spread to the world (Oppenheimer, 2012); however, recent discoveries related to the genome sequencing of archaic humans as Denisovan and Neanderthal and their comparison with human genome reveal that a much more complex peopling of the world, which involves admixture with these species, might have taken place (Gibson, 2011). According to "Out of Africa" humans left that continent in several migration waves about 60 to 120 thousand years ago, populating Europe, Asia, Oceania, and finally the American continent. Each wave consisted of a handful of individuals who carried with them only a small part of the original African genetic pool, so the genetic makeup of each wave was different from the others, which is referred to as "founder effect" (Li et al, 2008; von Cramon-Taubadel and Lycett, 2008). Continental serial founder effects, together with local forces as natural selection and genetic drift continued to act upon relatively isolated populations of wonderers. Isolation between two given populations is directly proportional to the geographic distance between them as this parameter decreases the magnitude of historical migrations (Marks et al., 2012). Although this is true for intracontinental populations, isolation between continents is higher because of natural barriers (Handley et al., 2007). Then, differentiation is a function of geographic seclusion, distance, and time. Continental characteristics appeared first, followed by variants within each continent, and finally variants specific to each community. During most of the time outside the African continent humanity devoted to populate the Earth and differentiate, as those first bacteria did for roughly two billion years.

With the advent of new forms of transportation that allowed traveling through hundreds and even thousands of kilometers, raids by armies and entire peoples were possible, conquering villages, absorbing and merging with them. The great empires of antiquity, such as Chinese, Roman and Assyrian, were multiethnic centers, breeding grounds for culture, invention and commerce (Deeg, 1999; Temin, 2001; Parpola, 2004). Trade relations were particularly important encouraging migration and with it, cultural and genetic admixture as well as the exchange of goods allowing the enjoyment of new products. One of the most famous and longest of such trades routes was the Silk Road which linked Asia and part of Africa to the Mediterranean and the rest of the European world since the third century BC (Comas et al., 1998; Liu, 2001). However, in the ancient world many commercial routes were formed, uniting kingdoms, empires and the towns of that time (De Navarro, 1925; Lathrap, 1973; Hirth, 1978).

The enrichment of the populations from the merging of two or more cultures, or simply with migration, led to many of the consequent cultural and technological advances. Genetic and cultural fusion has not stopped its course in our days, even exacerbating in the last century with the improvement of transportation and communication routes between towns and cities, as well as the increment of the demand of labor force in the latter. For instance, we have found a clear relation between the increment in magnitude of migration waves and the development of urban centers and roads when studying the Garifuna communities (Herrera-Paz et al., 2010). It is known that a good proportion of today's Europe is composed of African immigrants, and the world's most powerful country, the United States of America, is a composition of immigrants from all over the world (Schiller et al, 1992; Foner, 2000; Kohnert, 2007).

Perhaps the most notorious example of the effect of separation, differentiation, and reunion can be seen in modern technologies. Two researchers, or alternatively two teams of technologists or scientists, perhaps far apart, separated by time, physical distance, culture, their respective specialties, or even by the state of the art of science at that time, invent and build something. Then, unexpectedly, both technologies come together to form an entirely new, emerging one.

A mid-nineteenth century British mathematician named George Boole invented a type of algebra based on binary logic (Boole, 1847). Back then, nobody had the faintest idea of the possible practical applications of that Boolean algebra. At about the same time, a prestigious member of the Royal Society of London named Charles Babbage invented the first calculating machine (Swade and Babbage, 2001). Although Boole and Babbage were contemporaries and both members of the Royal Society their ideas kept separate until the mid-twentieth century, when merged and gave birth to the first computers that used the binary logic for processing information which later evolved to become essential elements of modern society. In fact, a modern computer is the product of the integration of hundreds, or perhaps thousands of small isolated ideas. Synergistic fusion of computers with other outstanding discoveries and inventions boosted scientific and technological advancements in the late 20th Century. The marriage of the computer with telecommunications originated the global network (the Internet), and the relationship of the computer with discoveries in genetics and molecular biology led to the sequencing of the human genome (Collins et al., 2004), which in turn was the first step towards the future control of our own biological evolution. Examples of emerging technological phenomena allowed by the encounter of two different technologies are the rule in our modern society and increasingly become more frequent, growing at an exponential rate.

Exponential Growth

In 1965 Gordon E. Moore (1998) published in *Electronics magazine* which claims to be one of the most popular assertions in modern geek society. Moore's law predicts that the number of components within integrated circuits in computers should double every 18 months. That is, every 18 months the amount of memory storage and processing speed double, and hence, the cost of production and the price of computers are greatly reduced. The doubling in computing power is produced in accordance with the miniaturization of components. And Moore's Law proved to be accurate! Moreover, the 18 months predicted by Moore has shrunk to 13 months in the last years (Kurzweil, 2003). If commercial aviation had experienced the same falling of prices as computers, one could travel anywhere in the world for a fraction of a dollar. The small portable unit that my daughter uses to listen to music, which can be found pretty much in any store in the world, has the storage capacity of the most powerful computer from ten years ago. I still remember the computer on which I learned BASIC and Pascal programming at the beginning of the nineties. It had a hard disk capacity of two Megabytes (MB), (expandable to four, the seller proudly told me). Today, some of the files containing high-resolution graphics stored in my laptop weight more than two Mb. And I have hundreds of those!

We say that the technology of digital circuits has grown exponentially, referring to an accelerated growth. It seems that each new development greatly facilitates the following. But the miniaturization of components has a limit, a restriction. At one point a component will become so small that it will start to follow the rules imposed by quantum mechanics, which are evident in the world of atomic and subatomic particles. Then, the binary digits will not be in a particular state of zero or one, but will start to fluctuate, located in a blur nebula between the two values following the rules of quantum superposition. Another concern is quantum tunneling, with electrons jumping from one wire to another. The quantum world is strange! Each particle can be in many places at once and a computer with circuits functioning in accordance to the quantum rules will become unstable.

So, in theory, Moore's Law should fail after a critical value of miniaturization. When achieved, the components cannot be further reduced in size and technology advancement will be stalled. That is why today many hardware companies in collusion with computer scientists and theoretical physicists work in the quantum computer (Ladd et al., 2010). The marriage of computer sciences with quantum theory may produce quantum computing, thereby making *in silica* integrated circuits technology obsolete. The computing power will be boosted, and not following Moore's law, but rising disruptively perhaps millions of times in the overnight. The storage and processing of information will have taken a big leap, but from this point, Moore's law will take control again.

The exponential growth of the advancement of technology is not limited to computers. A long way has been walked in the last decade in the study of the DNA. The sequencing of the human genome and its publication in 2003 opened a door to the new era of genomics. The project took long hours of work and collaboration of a large number of laboratories around the world, and its total cost was estimated in more than two billion US dollars. Once completed, the main goal was to determine variations in the sequences (human variome) that may result in genetic diseases, increased susceptibility to complex diseases, or simply in normal phenotypic polymorphisms (Kohonen-Corish et al., 2010; Lander, 2011). The discoveries are contributing to the development of new diagnostic methods and determination of pharmacologic targets for new treatments. Biologists and microbiologists, meanwhile, began to sequence the DNA of many species. All this resulted in progressive improvement and the rapid lowering of the coast of sequencing a complete human genome.

In 2007 Dr. Greig Venter — great contributor to the sequencing of the first human genome and a pioneer in synthetic biology — announced the complete sequencing of his own genome by his scientific staff at a price of US$70 million, a reduction of more than 28 times compared with the first genome (Levy et al., 2007). Later, James Watson's — co-discoverer, together with Francis Crick, of the double helix structure of DNA — genome was sequenced using second-generation technology at a cost of less than US$ 2 million, a price reduction of more than a thousand times (Wheeler et al., 2008). By the time of this writing (mid 2013) we have completed the DNA sequencing of more than 1,000 people of African descent through next-generation sequencing technology, including 50 Garifuna people of the Caribbean coast of Honduras, at a cost of around US$ 1,000 each (under the *Consortium on Asthma among African-Ancestry Populations in the Americas,* CAAPA). Price reduction with respect to the first genome has been nothing less than two million times in just a decade. However, by the time you read this maybe the price would have dropped to one hundred dollars per genome or even less, and the sequencing of a personal genome could be performed in any laboratory in any country.

The exponential growth of sequencing coupled with stem cell technology will solve many of the health problems that afflict humanity today. Soon, possibly within a few decades (but still, it could happen much sooner), human life expectancy will increase from around 80 years to perhaps 120, or even indefinitely. The theoretical physical immortality — theoretical because it will still be possible to die violently — however strange and romantic it may seem to us, is the direction toward which modern science points. The fountain of youth, the proverbial elixir of life, has been sought after by various cultures throughout history. It was one of the goals of the ancient alchemists, and in its search, Spanish conquistador Fernando Ponce de León found the current state of Florida, in the United States. Needless to say, the only thing immortalized from the conqueror was his name. Genetics, computers and stem cells are getting us closer each day to that invaluable treasure (two examples of recent developments and findings in regenerative medicine and genetics can be found in Guimaraes-Souza et al., 2012; and Beekman et al., 2013).

We have seen that evolution towards complexity rolls along gently for a while, but only to a certain point before the discovery or invention of a new technology causes a disruption; a big leap towards higher levels of complexity. Cell membrane allowed the initiation of cell life; endosymbiotic events in bacteria originated the eukaryotic cell; improved chemical intercellular communication aided in the rise of multicellular organisms, etc. Well, it turns out that the time between two of these jumps is gradually shorter, so we can make a gross generalization of Moore's law: the evolution towards complexity advances at an exponential rate.

Scale constraints

From my early childhood, I have a clear memory of a television series named "Land of the Giants". The persecution of the giants against the tiny stars of the series was unmerciful. I used to keep glued to the TV! That was because the stories of giants are always fascinating, and perhaps that is why they are included in many mythologies. The book of Genesis mentions the *Nephilim* or fallen children of human women with the "sons of God". Apparently, when the sons of God watched the exquisite lines of human women fell in love and married them, and begot these individuals who according to the Bible were true giants. The Bible finishes the passage affirming that "the same became mighty men which were of old, men of renown". Other giants in ancient literature include Titans and Cyclopes of the ancient Greeks, and the Frost Giants of Old Norse (Jakobsson, 2008; Clarke and Bolton, 2010).

About how tall should the *Nephilim* mentioned in the Bible or the mythology giants have been? Is their existence possible? Many believe the sons of God belong to pseudo human races coming from other planets; space travelers who liked to conquer new worlds. I do not know if the ancient stories of giants have some degree of truth, but what I can say without a doubt is that a being of more than 2.5 meters should have had a very different morphology compared to modern humans because of "scale constraints". This type of restrictions is seen in all levels of complexity. The evolution in size cannot continue to accumulate because of these constraints, so in turn, what increases is the number of relationships between individuals, and hence the size and complexity of the community.

Let's start with the case of a cell. Imagine a single-celled organism evolving in a beneficial environment with abundant nutrients. The number of organisms born in a generation is plentiful, so there is a stiff competition within the group. Larger cells will have some evolutionary advantage displacing smaller, maybe because of a higher mobility or some other positive characteristic related to size. Under this evolutionary pressure cells gradually begin to become larger as the generations pass. How long would these organisms grow in size? Is there a limit?

Most cells, including unicellular organisms, are tiny, microscopic creatures, not easily visible to the naked eye. Why didn't any cell evolve indefinitely to become as huge as humans, or perhaps dinosaurs or whales? Blame it on the scale constraints. When a size limit is reached any further evolution towards complexity will favor the next level. In a cell, the main restrictive mechanism of growth is given by its need to feed on the nutrients present in the environment, and also by the need to get rid of toxic substances produced by metabolism (interesting revisions of other phenotypic constraints, including some that aid in cooperation and complexity, can be found in Foster et al., 2004; Wagner, 2011; Berkhout et al., 2013). Now, suppose we have a single-celled organism close to that evolutionary size limit. Suddenly, a genetic mutation that increases the diameter of one of the daughter cells to two times the original appears. What would happen next?

To begin with, the cell volume increases in a cubic relationship with increment of diameter. For instance, if the progenitor cell measures two micrometers in diameter, then its volume is approximately 4.19 cubic micrometers (using the equation of the volume of a sphere). The mutated daughter will measure then four micrometers in diameter (twice), and the volume will be about 33.49 cubic micrometers. That is, its volume increased eight times over its predecessor because it grew in a cubic relationship. Nutritional demands of the cell and the amount of waste produced is volume dependent, therefore these parameters also increase eight times compared to its mother. The daughter cell will need eight times more food and produce eight times more waste.

In the other hand, nutrient uptake and excretion of waste are processes occurring across the cell membrane. But the membrane is a surface and as such, its growth increases in a square (not cubic) relationship with diameter. In the case that concerns us the membrane of the not mutated progenitor cell will have an area of 12.56 square microns, and the mutated daughter cell will have an area of 50.24 square micrometers. That is, the exchange area will have grown only four times, while the metabolic demands and the production of waste are eight times higher.

If the cell continues to grow two things should happen: 1) the rate of absorption of nutrients will be insufficient for the metabolic demand. 2) The rate of elimination of toxic substances will be insufficient to the amount produced. As a result the mutated cell will most probably die by starvation and poisoning, eliminating the mutated gene from the population gene pool. Nothing more convenient for a cell — considered as an evolutionary entity — than to take advantage of energy resources, not for further growth in size, but to specialize in order to help their neighbors while benefiting from the output produced by their peers contributing to the growth of the population, thereby increasing the complexity of the next level.

What about a human being? What if he (or she) grew up, for instance, more than three meters tall? In this case, the weight of an individual is given in proportion to the volume, growing in an approximate cubic relationship with increasing height. However, the muscular strength of a limb (a leg for example) depends on the cross sectional area of the muscle. An individual more than 3 meters tall would be extremely heavy to hold with a pair of legs of normal proportions. Limbs must then grow to be disproportionately thick in relation to the trunk. Additionally, among mammals as body size increases brain size increases in a negatively allometric manner, following a power function with an exponent of 0.6–0.8 (Roth and Dicke, 2005); then, the head would be disproportionately small with respect to the rest of the body. But then again weight rises in a cubic relation, and the strength of

86

the neck muscles in a square relation. Consequently, the neck has to become extremely thick to sustain a heavier head. For that reason mosquitoes can have very thin necks and legs in relation to the rest of their bodies, while elephants have them wide. And the larger the animal the more noticeable will the gravitational effects be on its economy and body shape. A mosquito of the size of an elephant would die crushed by its own weight just before its head had fallen and lapped.

And muscle strength is only one of the multiple constraints in maximum growth of an animal. An important issue, at least among endothermic (warm-blooded) animals, is the dissipation of the heat produced by metabolism. The heat production increases in proportion to the volume in a cubic relation, whether the heat dissipation capacity increases in proportion to the skin area (Phillips and Heath, 1995). And that's why elephants have such large ears which increase their cooling area! When I was a little kid I thought, wrongly, that the only purpose of elephants' large ears was to get rid of flies.

It is obvious from the above that there is a limit or restriction on the potential size of an individual, and so the things, I cannot imagine how an alien visitor conqueror of multiple worlds could seduce a female human earthling. I can hardly believe that a giant of more than three meters tall with thick limbs and neck, walking slowly and making the ground beneath him vibrate with each step, would had ever been attractive to picky earthling women.

Therefore, the existence of a group of celestial giants traveling from galaxy to galaxy, conquering every female of every habitable planet they found, and also mating with them, is unlikely (though not impossible). Similarly, it is unlikely that a single gigantic supercomputer would ever control the flow of information and energy processes of a planet or a galaxy.

And that's precisely the theme of Isaac Asimov's brilliant essay called "The Last Question" (Asimov, 1956). The main character is a giant supercomputer that runs the universe. In 2061 Multivac (this is the name given by Asimov to the computer) managed all global resources. As the thousands, millions, and finally billions of years passed, the computer (then called AC) had grown in size and ran all the resources of the solar system, the galaxy and then finally the cosmos. In the end, the merger of the universal human mind and the cosmic supercomputer regenerates the universe after the heat death caused by entropy.

The manuscript from the famous science fiction writer is nothing less than brilliant and amazing, but by 1950s computers called mainframes were giants that promised to continue growing in size to control increasingly complex processes, so it is not surprising that Asimov adopted this configuration for his science fiction essay.

A company opted for a monopoly on mainframes becoming one of the most powerful financial empires of its time (during the

1950s IBM 700/7000 series dominated the market). By the decade of the seventies, International Business Machines (IBM) had grown at the same rate as their mainframes becoming the absolute giant of computers (Bashe et al., 1986; Campbell-Kelly et al., 2004). Then, two young men built in a garage what would be the future: the personal computer. Steve Jobs and Steve Wozniak sold primitive and small computing devices assembled in a garage to their neighbors and, well, you know the story. One of the pillars of the transition from the industrial age to the information age emerged, as well as one of the most profitable technology companies in history (Imbimbo, 2009). Soon, most people around the planet will have a computer connected to the global network. The giant dinosaur of Asimov's essay was replaced by a network of well connected ants which will someday manage the planet's resources. In computers as in humans, complexity is rapidly accumulating to the group level.

Entropy

Shortly after the beginning of this essay I told you that there are two opposing tendencies in the universe, and I am returning to this issue because of its importance in the understanding of evolution towards complexity. There is a trend determined by the three forces organizing matter into levels of increasing complexity. The other is destructive and is called entropy. Entropy leads to thermal equilibrium of the universe, disrupting the organized in such a way that makes everything evenly distributed throughout space. Entropy guides the physical universe to a state of maximum randomness, diminishing the amount of usable energy. So it is destructive. And it may act very fast! Entropy can destroy in a few seconds what was built by the three forces during millions of years.

Then, how is it that life evolves into increasingly complex levels despite entropy? Strong arguments have been put forward to explain this dilemma. Biologists and physicists explain that in a non-adiabatic open system it is possible to increase local order (life on Earth), provided an increase in the overall disorder (in the universe). That is, life on Earth is possible because the amount of energy supplied by the sun is huge, but only a small part of it is used to generate order and complexity. The rest is dissipated as heat (disorder) so that global disorder always increases (Schneider and Kay, 1995). The above is a half truth. While that explanation satisfies the second law of thermodynamics (Schneider and Kay, 1994) it is also true that it tells us nothing about the role of entropy within the system evolving towards complexity. Physicists that have developed the theory of complex systems tell us that in an open system in disequilibrium, with high energy input, the components exhibit a phenomenon called self-organization by which coordinated behaviors and emergent properties arise, but the role of entropy in the occurrence of this phenomenon is not entirely clear (Nicolis and Prigogine, 1971; Ge and Qian, 2011). In the following lines I'll explain why entropy is as important—in order to produce complexity—as the three forces of nature, at least in living systems and their derivatives.

In 1859 British naturalist Charles Darwin published his seminal and revolutionary "The Origin of Species by Means of Natural Selection" in which he explained the way individuals with favorable characteristics are more likely to survive, inheriting their features to offspring (Darwin, 2009). Despite being contemporary of a man regarded as the father of genetics, Darwin never met a Czech monk named Gregor Mendel or his laws of inheritance. Darwin never knew or even suspected the source of variation, and it took several decades for his theory to be complemented with the knowledge of genetic variability and the fusion of both into a modern synthesis (Kutschera and Niklas, 2004).

Imagine a population from a certain species. If all individuals were identical the population would not evolve, so the introduction of some variation becomes indispensable. In living organisms, it is still considered that mutations are the primary source of variability at the genetic level (Gommans et al., 2009). A mutation is a disruptive, destructive event, product of entropy. Mutations are random and tend to destroy the genetic information that needed millions of years to accumulate. But variability is necessary because occasionally a mutation that increases the information content occurs, improving a characteristic of the carrier individual making it more adapted to its environment. The frequency with which these beneficial mutations happen is low, but apparently sufficient for the mechanism of natural selection to drive the population toward adaptation (Peck, 1994; Oor, 2010).

Cells count with potent repairing mechanisms to ensure the fidelity of DNA copying during replication prior to cell divisions in gametogenesis. Nonetheless, it seems evolution has allowed certain degree of imperfection in these mechanisms so variability can take place. For me, this is a stupendous example of group selection since deleterious mutations exceed in number to the beneficial ones, but variability is favorable for the group in order to adapt to ever changing environments and evolve. Repairing mechanisms must be fine-tuned to allow mutations.

Although natural selection is clearly and elegantly explained in the work of Darwin, it is not the only mechanism that drives evolution toward higher complexity. In fact, the theory of evolution of Darwin and Wallace as we know it provides few clues about why life grows into increasingly complex forms, and not vice versa. While—as previously mentioned—life in community may offer selective advantages over solitary individuals, I argue that these benefits are necessary but not sufficient to explain the apparent direction of evolution towards more complex forms. If genetic advantages of gregarious individuals over solitary were strong enough to drive, *per se*, evolution to complexity, then, every level of complexity might have been absorbed by higher levels, and this is not the case. A species of amoeba may be as well adapted to its environment as the most complex of human societies. All levels of complexity from viruses to complex communities coexist in planet Earth. So, if an evolutionary advantage in complex over simpler forms exists it may be too small to explain the actual degree of complexity of social insects, human populations and ecosystems, just to mention a few examples. Then, we will have to dig deeper, or look somewhere else.

The answer to the above dilemma is simple. Indeed, mutations that are not beneficial, those that break down or decrease part or all of the function of genes —for the purpose of this discussion, gene silencing in the case of cells or any type of lost or silencing of information at the individual level would be equivalents—favor the growth in complexity. In other words, the increase in entropy or disorder in a level favors an increment in complexity in the next upper level. I shall call it "simplifying entropy". Although deleterious mutations with subsequent loss of function has been amply studied in the context of genetic diseases in humans (Li et al., 2010; MacArthur et al., 2011; Marian, 2013), accumulation of mutations and other silencing events acting as simplifying processes may have a broader influence in augmenting complexity and its importance has been underestimated. I will take some time to explain this, and I shall do it with examples. In these, I'll introduce elements that I think are fundamental to this theory.

International Shoes Inc. The way in which a small artisan business grows into a large firm of mass production is a superb example of the direction from simple to complex. At the time of its foundation, "Herrera Sandals" was a small shoe manufacturing company that hired five artisan shoemakers who carefully manufactured, each, five pairs of shoes a day, which were then sold at a relatively expensive price only to the wealthiest of people from town.

The owner of the company decided to make a small investment in his employees, and sent one by one to specialize in a particular task of the manufacturing process. Once employees specialized, the owner discovered that they could make many more shoes in one day, as each one was focused on a single task and performed its labor faster. This reduced the price of the pair, so considerably more people could buy shoes in town. The rise in sales enabled to hire more skilled employees, thus allowing a surplus of manufactured shoes which were exported, and thus "Herrera Sandals " became " International Shoes Inc".

The first element in this story is the capital. Capital in human society is the equivalent to the energy in biological organisms. If we draw the line from the capital represented in any form to its source, we see that inevitably ends up in the sun's radiant energy. All other factors being equal, the price of a product or service is roughly proportional to the energy that has been currently and historically invested in its production. As the amount of energy needed for making a pair of shoes gradually lowers, the price will be lower and more people can put on shoes (the above mentioned economy of scales).

The second element is the specialization of employees. As the employee gains expertise in a specific field of production, he actually becomes simplified since he now only needs the knowledge of a small part of the process and not of all the five stages. Never again will any employee of International Shoes Inc. waste his valuable time learning the skills necessary at all stages of production, so we got to the third element: Simplification.

The fourth element is the increased interdependence. A particular employee cannot make a single pair of shoes, so all of them depend upon everyone else to bring the manufacturing process to fruition (and thus keep their jobs).

If we put together all these elements the irreversibility becomes evident, and this is common to every evolutionary level. I will illustrate this fact joining two pieces of this puzzle: Capital and simplification. In the first place, energy expenditure (capital) that would be necessary to re-train the skilled workers in all stages of the manufacturing process would be far greater than that needed for them to specialize in a single task, so that kind of training is not only unnecessary but very costly and unattainable. In second place, each expert is far more profitable than the craftsman who knows how to do everything. But indeed, is far simpler than the craftsman. This fact makes the mass production company be hopelessly driven towards greater complexity, without alternative–assuming that all necessary resources for the manufacturing process are present, together with a high demand of the product. We thus see that the upper level (the company) grows in complexity as elements in the level below (employees) specialize but get simplified in some or many ways.

I realize that you would probably be skeptical on my affirmation that irreversibility of evolution towards complexity is (in a great part) imposed by entropy, so I have a few more examples to show you.

Near-sighted hunters and big-headed babies. Consider a cell in our body, specialized in one particular task. Because in order to make its specific work the cell needs only a handful of genes, most of the remainder are typically silenced (Vaissière et al., 2008). This process was thought to be largely irreversible, although now it is possible to reverse it in the laboratory (Park et al., 2007). The specialized cell is much simpler than a unicellular protist like, say, yeast, which must synthesize all the proteins needed on its own to survive (although its genetic expression profile changes with each stage of life cycle and environmental shifts; Chu et al., 1998). The specialized cell of our body, however, depends on the supply of many other cell types in our organism. The increased complexity of the human organism is accompanied by the simplification of its cellular components. The process is irreversible since it is much easier and cheaper to silence genes, than to activate all of them.

The cell would not tolerate the energy demand of synthesizing every single protein. Besides, why should the cell do that? Everything it needs beyond its own production is supplied by its peers or by the overall organism.

Here is another example. I am shortsighted. When I walk each morning into my physiology class, it amazes me that almost half of my students wear glasses for nearsightedness. Then my mind goes back some 20,000 years ago when human hordes of hunters and gatherers roamed European and Asian landscapes looking for the best sites for mammoth hunting (Germonpré et al., 2008). No doubt being shortsighted in those days was a huge disadvantage. For sure, because they probably fumble when hunting, many poor shortsighted hunters were condemned to ostracism, and with it, their ability to reproduce was minimized.

Notwithstanding that adverse allelic variants that confer susceptibility to myopia (Jacobi and Pusch, 2010) might have been appearing since long time ago, its occurrence in human populations should have been kept low thanks to this type of negative selection. With the advent of agriculture, sedentarism, division of labor and specialization a number of occupations that no longer needed a perfect sight emerged. Moreover, the recent appearance on the scene of particular types of specialists, optometrists and ophthalmologists, allowed myopic individuals to have a normal vision by the use of lenses. The pressure of natural selection ceased, so the new genetic variants originated by disruptive mutations of vision genes could survive and be transmitted to the next generations, spreading in the population. The result is a high proportion of myopic people today (Saw et al., 2010). Natural selection no longer acts against us, and the use of glasses even gives us an intellectual look!

Specialization, primarily in the medical branches, has led to a simplification of human characteristics with respect to our ancestors who lived thousands of years ago. Individuals from that time were required to master a wide range of activities for their survival, including the construction of their own homes, and making their own clothing and weaponry. Before 12,000 to 10,000 years ago human societies had no specialist (Massey, 2002). In turn, immune systems would have been very strong in those environments of little healthcare. The emergence of individuals specializing in healing and public health may allow some unfavorable genetic variants, with the proper care, to survive and to be passed on to offspring and spread in the population. The use of antibiotics, for example, may permit mutations (and thus entropy) that decrease the efficiency of the many genes that comprise the immune system to survive, which might determine, in a few generations, an increased percentage of weak immune systems with poor capacity to defend us from bacterial infections. But who will need strong immune systems anyway? We will have potent bactericide substances.

We see how simplification (weaker immune systems) goes

hand in hand with specialization (in this case, pharmacology and medicine) which in turn makes us increasingly interdependent, but that contributes to the increasing complexity of societies. And the same is true for the rest of the tremendous arsenal with which modern medicine takes care of us, from vaccines to grafts, just to name an area of knowledge. At the same time the number of occupations and labor niches gradually increases lowering the proportion of the contribution of any single person to a community. In fact, the work that each of us does is minuscule compared to the total activity of society. This immediately suggests us a way to measure the degree of complexity in a society or in any other level: Simply, complexity of a level within the system will be, roughly, inversely proportional to the mean percentage of the total labor done by single individuals or elements, or equivalently, directly proportional to the number of niches. For instance, the number of different cell types of an organism has already been used as a measure of the complexity of a metazoarian (Arendt, 2008).

The third example has to do with the very nature of human beings and is an aspect that differentiates us from our ape cousins. Females of the great apes give birth with low pain, and delivery is generally easy and self-assisted (although there is some recent evidence against; Irata et al., 2011). However in humans the large size of the child's brain, and thus the diameter of the head, predisposes to difficult delivery which requires the assistance of other persons (Wittman and Wall, 2007). As the process of encephalization occurred in proto-human societies individuals who specialized in assisting deliveries appeared; perhaps relatives and other females from the community at those prehistoric times, and midwives and doctors in the present (Rosenberg and Trevathan, 2002). The emergence of these specialists helped reduce mortality in children with large heads, but made us dependent upon specialization.

Did encephalization stopped? Or is it possible that mean measurements of the human head continue to grow in our days? And if this is the case, under what kind of evolutionary pressure are they still growing? Comparisons of cephalometric measurements between skulls from people who lived in the 14th, 16th and 20th centuries have shown a significant and progressive increment in forehead (and hence brain) size, together with significant simplification of facial traits (Rock et al., 2006). Child and maternal death rates during the medieval ages where extremely high until the advent of medical advancements such as anesthesia, transfusions, asepsis, cesarean operation, and more recently, antibiotics (Wells, 1975; Todman, 2007; Wells et al., 2012). In the present, one of the main causes of cesarean operation in hospitals around the world is a head that does not fit into the birth canal, namely cephalopelvic disproportion with the consequent obstructed labor (Gaym, 2002; Dafallah et al, 2003). But today the size of the baby's head is not a big concern for the survival of both the mother and the infant. Mortality rates for this cause have dropped substantially, at least in those places with adequate health care services. However, maternal and child death must have been important evolutionary constraints during the middle ages and before, subjecting genetic variants that contributed to large head size in newborns and narrow pelvis diameters in mothers to a strong negative selection (Grabowski, 2013).

Is it possible that improvement in obstetric and medical procedures and cesarean operation have been acting as factors for the modification of this trait (head size) in the last centuries? Has modern medicine been the main factor contributing to the size increment of the human forehead described in Rock et al., 2006? If this is true, nonetheless there might be no positive selection over intelligence today, it could continue rising because of relaxation of an evolutionary constraint, i.e., a selective force ceased to act upon the maintenance of head size under a certain limit. Moreover, the tendency towards subsequent head, brain and intelligence growth should continue in the following centuries — that is, assuming that intelligence directly correlates with increment in brain and cranial size, which has not been entirely proved.

While it might be true that larger heads associate with larger and more complex and intelligent humans, on the other hand from the standpoint of reproduction we are simpler since our females are less able to bring to the world a new person without help. Moreover, evolutionary relaxation might allow women with narrow pelvic diameters to survive delivery thanks to cesarean operation, and so their children, receptors of genetic variants for narrow pelvises, which in turn might increase the proportion of cesarean operations due to cephalopelvic disproportion in the future, which in turn might allow delivery of children with bigger heads. This circle could continue—although very slightly and even unnoticed—from generation to generation until other constraints are met (loss of sexual attractiveness, for instance). Dependence upon modern medicine at the time of childbirth greatly increases social interdependence contributing to the complexity of human society.

The next example has to do with education, the core of civilization. In ancient times, before the invention of printing, books existed as singular copies that in most cases were considered invaluable. When a wise man or a scholar had access to a book he (or she) used to memorize it, so, transmission of knowledge strongly relied upon memory (Mckitterick, 2000). Then, with the advent of the printing press it was not necessary to do that anymore since it is always relatively easy to re-find a book if the information is needed. Besides, why spend valuable time memorizing a single book when many are printed? Just grasp the essential ideas. But even this practice is becoming obsolete in our times of rapid progress. As example, a few decades ago a biochemistry student had to study a few metabolic pathways, but at present, this discipline has experienced a remarkable growth and it is no longer possible to memorize the thousands of metabolic pathways, small molecules and molecular interactions discovered (Pearson, 2007; Vidal et al., 2011). Furthermore, massive data storage and devices designed to keep humans online make universal information (among which are biochemical substances and pathways) permanently available in the blink of an eye.

What is the point in memorizing much of a textbook when information can be obtained almost instantly? But then, specialization in only a small fraction of knowledge becomes necessary, and that is why productive work in science is, today, the result of complex international and multidisciplinary networks (Wagner and Leydesdorff, 2005). The solitary scientist, the Lone Ranger, the polymath genius so popular at the dawn of science, loses its practicality in the complex society. On the other hand if we do not need to memorize much then those genes involved in memory, perhaps extremely important in ancient times, can relax a bit and a few random mutations can accumulate from generation to generation and disperse in the population. Who knows? We might be getting simpler in that sense (memorization), and at the same time, that could make us gradually more dependent upon the ubiquitous universal memory provided by the network.

Brains *à la Carte*. Our last example is also related to evolution of human brains. Today, genetic evolution can be assessed by comparing changes in whole genomes of two related species. The sequencing of the human and chimpanzee genomes a few years ago has enabled their comparison. It has been found that both are extremely similar, with no evidence that the activation of novel genes has been an important mechanism in the development of the human brain (Hill and Walsh, 2005). Moreover, many of the genes found in the chimpanzee are inactivated in humans (they are now pseudogenes), especially those corresponding to olfactory receptors, (Gilad et al., 2005) which is evidence of some simplification in humans with respect to chimps.

The differential evolution of a gene can be evaluated by means of nucleotide substitutions between both species. If the number of synonymous mutations—i.e. the ones that do not change the aminoacid in the resulting protein—is much greater that the non-synonymous, the gene is said to be highly conserved and most historical mutations has been deleterious, hence suffering negative selection (though there is some evidence that challenges this assumption; Chamary et al., 2006). In the other extreme, if non-synonymous mutations are predominant, they have most probably experienced a positive selection. Evidence is compatible with the fact that many genes involved in neural networks in humans are highly conserved except for a handful of them that has undergone very strong positive selection—the aforesaid *FOXP2* related to speech is one of them. For me, it is outstanding that out of many, only a very small number of cerebral genes specialized in the process of transforming a primate brain into human.

If simplification of brains originating interdependence into some extent has been an important additional factor in the development of the social human, we would have to take a better look to neural genes with high numbers of both synonymous and non-synonymous mutations, which are usually considered as neutral. These loci would not be neutral in chimpanzees and might be highly conserved among that species, but evolutionary relaxation due to cultural aids would approximate them to neutrality in our own species permitting mutations to accumulate, consequently raising the number of allelic variants. If the theory I present here is correct, the number of conserved genes in chimps that are somehow relaxed in humans would be higher that the opposite, not because of intelligence, but because of our increased tendency to group cohesion and interdependence. Sadly, simplification as an element to complexity is not addressed in literature with the importance it deserves, and comparison of the proportion of neutral or quasi-neutral DNA sites between humans and chimpanzees may have to wait until many more genomes of the latter are sequenced. Putting it in simple words, natural selection must have a much stronger impact in chimpanzees compared to humans due to social complexity.

But if there is still a long way to go in the study of evolution of human brains, we have at least one amply studied set of traits that experienced enormous simplification due to a cultural invention: the masticatory complex and the guts. Chimpanzee food consisting of slightly bitter, high fiber fruits may result disgusting for humans. One would barely survive with such a diet, unless, of course, you are a skinny model (you won't eat anyway). Shortly after the discovery of fire the invention of cooking could have shrunk our ancestor's teeth and made their inferior maxillary and intestines smaller. Cooking makes food softer, with less fiber and more digestible which in turn improves calorie intake, something necessary for a growing brain with a high metabolism and hence, a high caloric demand. Additionally, the combination of ingredients in the preparation of dishes requires ingenuity. As brains grew, most components of the digestive system shrunk. A high positive correlation between diet quality (that allows intestinal simplification) and brain size in primates has been demonstrated. According to the Expensive Tissue Hypothesis simplification of the intestinal tract is a trade-off. As cerebral volume increases, the body experiences an energy crisis whose consequences are suffered by intestines (Fish and Lockwood, 2003).

Cooking and human eusociality probably co-evolved (Wrangham and Conklin-Brittain, 2003; Driver, 2010). It is relatively easy for me—I guess, nonetheless this hypothesis has not been proved yet—to make a fire and cook a couple of eggs. After all, I have matches, which have been made by many people in collaboration. But when we go camping and there are no matches around, I have a huge problem because I am incapable of lighting a campfire on my own; but can still ask the nearest Boy Scout for help. Sure I can eat raw chimpanzee food, however, at the risk of watching my intestines explode. I am simple and dependent. But I don't starve because I am eusocial. What a great advantage! (Nowak et al., 2010 beautifully address the evolution of eusociality in insects by means of group selection; however, its frontal confrontation with Inclusive Fitness theory has unleashed controversy: Abbot et al., 2011).

These examples allow us to infer that the probability of survival of a human outside the protective sphere of society is lower today than ever before since we are simpler and more dependent. We just have to remember the stories of castaways depicted by Hollywood movies to realize how difficult for a human being is to survive on his or her own. The increased complexity of communities (increased number of relationships between individuals) is paired with simplification, specialization and interdependence of individuals.

Now imagine a country ruled by a tyrant, enemy of progress. By decree, the tyrant orders to reverse back to ancient times where everyone planted, grew and harvested their own vegetables, raised their own cattle, manufactured their clothing, constructed their own houses, attended their own medical needs, etc. No doubt this society would collapse because we have lost the ability to do all those things. Modern manufacturing processes are so specialized that there is no single person in the world capable of fabricating even a modest pencil by his or her own (Read, 1958). Progress, an expression of increasing complexity, is mostly irreversible since it is impossible to train and reprogram our genomes and our brains so that we all do everything. The energy cost would be just too high. Simplification is loss of information at the individual level, genetic or cultural, and is the product of entropy. Its reversal is extremely difficult. And what is true for humans, is also true for cells that make up a multicellular organism, or molecules that make up a cell. The events leading to complexity are repeated at each level, like fractals, and organisms tend to evolve to levels of increasing complexity.

Of course, it is always possible to destroy a system or a huge part of it. If our tyrant razed 99.99% of the population, many genetic variants that confer specific skills and much part of the cultural and technological development would not be represented in the remaining 0.01%. A vast amount of biological and cultural data would vanish. Under these conditions of information loss in the overall population people would be forced to work harder, only the most versatile individuals would survive and society as a whole would be greatly simplified. Destroying information in the overall system is a way to stall evolution (Spielman et al., 2004) and reverse the progression towards complexity. Although biological systems are resilient, that is, has been shaped by evolution to bear with destruction (Holling, 1973), ones a threshold is reached information is irretrievably lost.

This leads us to a surprising conclusion that well deserves to be raised to the level of rule — although rule of thumb, I must say:

the evolution towards complexity in a biological system roughly depends on the interaction between the forces that tend to the destruction of the system as a whole, like excessive predation, unmerciful hunting, lack of food, meteorites, epidemics, atomic bombs, and genocidal tyrants, and the ones that leads to simplification of the elements, such as mutations, gene silencing or choosing a specific occupation. Small doses of destruction—environmental pressure—could aid in the activation of mechanisms of complexity at the group level, but beyond a threshold these are halted, or even reversed. In these conditions of increased destruction and stress complexity in the individual level should be favored. For me, it makes no sense to be a simplified specialist if other specialists are most probably already dead. Who will give me what I need? I have to take care of my own.

In general, we humans are programmed to see destruction as negative and evil, associated with the "the dark side". This perception of reality is necessary because to increase our probabilities of survival in society we must produce order and complexity through our work and through our relations with others. But we also know that sometimes we need destruction, or at least we think it's necessary. We destroy the lives of animals and plants for food, we destroy old buildings to build new ones, and destroy entire cities in wars. From the evolutionary point of view destruction has a broader implication.

Destruction and renewal. The race of life forms towards

complexity is like a maze full of obstacles. The different forms of life run through the maze trying to survive, changing their ways to overcome the obstacles, which represent selective forces. Occasionally, an organism enters in an evolutionary dead end and becomes unable to continue evolving, giving up to the destructive forces of the environment.

Since its emergence on Earth life has taken many forms, but most of them failed to pass the test and became extinct. Descendents of those that make it through the maze pass to another maze: the next level of complexity. Some life forms are in the lead in the race to complexity, as us and eusocial insects, but many others have remained in one of the lower levels, adapting wonderfully. In this race destruction ensures that only the most adapted survive. Annihilation resides at the very basis of natural selection and evolution (Raup, 1986).

Within each eukaryotic cell, the production of specific proteins must be very well regulated through various types of mechanisms in order to satisfy the requirements of the organism, just like the supply of a specific good or service corresponds to its demand in a free market system. The regulation takes place at several levels, and some of the steps involve destruction. Messenger RNA molecules are like photocopies (transcripts) of the protein blueprints (genes) located at the information central (cell nucleus). The messenger leaves the nucleus into the cytoplasm where is then read by the translational machinery (ribosome and enzymes) in order to construct a protein in a fashion that reminds us a production line. One messenger can be used to construct many protein molecules, so, in order to stop the production (when necessary) the messenger RNA strands have to be destroyed. Because demand of specific proteins can change from one moment to another and anomalous proteins can be synthesized from deleterious somatic mutations, permanent mechanisms of RNA destruction become necessary (Byers, 2002; Tijsterman, 2002; Shyu et al., 2008). The destruction of these molecules within the cell is massive and unmerciful. Before they are used to synthesize protein, only a fraction of the messengers produced within the cell nucleus survives. The life expectancy of a given messenger once it reaches the cytoplasm is just of several minutes. Messenger "assassins" called nucleases are so abundant in nature that their presence is an obstacle for RNA laboratory studies. At first sight, degradation of information carriers seems to be an enormous waste of energy; nevertheless it is extremely necessary for the cell and the organism in general to adjust rapidly to changes in protein requirement and to perform an efficient quality control.

Meanwhile, most intracellular proteins are synthesized, perform their function for some time, denature (grow old), and are finally destroyed (death) in the cell's junkyard called proteasome, all in a continuous process. In turn, old extracellular proteins are "swallowed" and destroyed by scavenger cells called macrophages (Glickman and Ciechanover, 2002; Ciechanover,

2005). Destruction guarantees that the amounts are appropriate at all times, and that aberrant useless specimens are properly eliminated. Aging and death are clear examples of the need for destruction in order for the group to adapt to changing environments (Longo et al., 2005).

Every human lives again, literally, all previous evolutionary stages through which humanity has passed. We begin our life as a single-celled organism (zygote), and each of the stages of differentiation of the embryo (ontogeny) simulates a stage of evolutionary development (phylogeny) in a process referred to as recapitulation (Evo-Devo; Gilbert et al., 1996). In this representation many anatomical structures that in many cases serve as substrates for other structures—but in others do not have any apparent function—are formed and then disappear. They are destroyed in a kind of self-annihilation of cells called apoptosis or programmed cell death. Many structures resemble those that once existed and had a utility in a distant predecessor but were not necessary for survival in later evolutionary stages. Cell death taking place in the growing embryo while reviewing the evolutionary history of the species, is impressive (Cole and Ross, 2001; Gjørret et al., 2003). Modeling of a human being's body beginning with a zygote resembles evolution of our species from an ancient unicellular eukaryote (Shumway, 1932), and throughout both processes destruction plays an important role. Most of our predecessor species have been lost in the mists of time, never to return.

One organ with a high rate of cell death is the brain (Kaplan and Miller, 2000; Sastry and Rao, 2000; Yuan and Yankner, 2000). Neurons are produced in excess, and many undergo apoptosis every day in the embryo and fetus, and even though destruction rate drastically decreases with growth and aging, it is always maintained for the rest of life. Neuronal death facilitates shaping and reshaping of synaptic relations and the consequent formation of different configurations of neural networks in a process mediated by neurotrophines and transcription factors, which in turn allows rapid learning and adaptation. This process has been called "neural plasticity" (Gutierrez and Davies, 2011).

Most cells composing a human body divide continuously. In each division cycle the strands of DNA must be faithfully copied, and this is done quite well for the first (approximately) 30 years of life by the DNA repairing machinery. But around 30 years old genes that codify for proteins that form this machinery have already suffered enough mutations to begin to weaken. From that moment aging gradually becomes more apparent in a person due to somatic mutations, mainly in mitochondrial DNA. Mutations slowly accumulate in different cell types including stem cells, and cellular senescence continues to increase until organs begin to fail and the individual eventually dies (Norddahl et al., 2011; Kennedy et al., 2012; Vijg and Suh, 2013). Apparently, evolution has favored the aging process in populations so that new generations take over and dominate. Destruction and death make sense in complex living systems, and they have a purpose:

ensuring the adaptation to changing environments with new variants replacing the old and damaged ones, and the same stands for a protein, a cell or a human.

EO Wilson, an American entomologist and expert in the study of insect communities, compares populations of ants and humans trying to establish as why eusociality in living structures emerges. In his latest book, Wilson (2012) argues that even though selection of genes that lead to selfish behavior is beneficial to the human individual giving him or her advantage over his peers, it is also true that cooperation and teamwork favors the group, so, altruistic genes prevail due to group selection. For thousands of years our ancestors lived their lives as nomadic hunter-gatherers conforming groups of 30 to 40 people, at most. This was an era of fierce competition between tribes for the resources of territories. In this sense, war has been a major factor for evolution of eusociality in humans, selecting genes that predispose to cooperation and compromise within groups. We see traces of these genes at present in the emotions aroused when our favorite team wins a game, or in the loyalty professed to our associations and ideologies. If that is true one of the main contributors to the genetic background of our eusociality would have been warfare. The need for cooperation within groups upon competition among them led us to the head of evolution towards complexity.

Dr. Wilson's theory deserves a comment. The historical influence of competition and destruction of foreign human groups by means of warfare to make us what we are is neither a license nor an excuse for beginning new wars, to support xenophobic and eugenic ideologies, or to continue with our destructive trends (the impact of human beings in the biosphere and other planet resources has marked a new epoch referred to as "Anthropocene"; Smith and Zeder, 2013). We have reached a defining moment. We have the technology to perform massive genocides and destruction of our environment, but also to positively change the environment and genes through technology, without harm to others. New technologies make military confrontations unnecessary for our subsequent evolution toward higher complexity. Rather belligerence, extreme selfishness and aggressiveness have become liabilities, vestigial behaviors; remnants of our evolution that today threaten us with our own extinction.

And then again at this moment as in every other in history our civilization can take one of two routes: the path to destruction or the path to evolution to the next level of complexity. Humans today or at any other moment, and like any other living being in this world, live in the razor's edge building complexity but always on the verge of destruction. Putting it in the language of complex systems, humans and human societies are dissipative structures at permanent risk of experiencing self-organized criticality events (that is, every once in a while a system evolving to complexity accumulates sufficient tension to experience a spontaneous collapse in the form of an avalanche, returning the system to equilibrium. The time lapse between two of such avalanches follows a power-law, with smaller events occurring more often than huge ones: Prigogine and Lefever, 1968; Bak et al., 1987; Perry, 1995; Sole et al., 1997; Brunk, 2002).

Two sequences toward complexity

Differentiation (radiation) and cooperation within the group. As a summary of the above, I would say that there are at least two basic sequences of events in the generation of complexity in living creatures. In the first one, we initially have more or less homogeneous groups of beings competing with each other. Groups consisting of cooperative individuals may have an advantage over others, so genes or memes for cooperation are selected. But the triumph and survival of the cooperating group is just the beginning. Mutations, small educational and cultural variations within the group, or entropy in general, as the case, continue causing variability so that the division of labor will begin to take place. Under these conditions specific subgroups performing specialized tasks begin to form. The remaining genetic information (or cultural or any other kind of information) intended to perform other tasks, will no longer be too necessary and random mutations or any other equivalent process of information silencing will start accumulating in individuals making them progressively more simple and interdependent, but specialized. In this mechanism, the trigger is the strong selection of cooperative groups, and from then on specialization and

116

simplification of individuals guided by entropy will be, progressively, mutually reinforced in a positive feedback loop. A good example is the aforesaid evolution of eusociality in humans.

Merging of two differentiated groups. The second sequence involves separation and reunion. A group is divided into two, A and B, isolated groups. The subgroups are kept separated from each other. During separation time random mutations, natural selection and other forces will tend to differentiate one from another progressively. After a variable amount of time which can be of hundreds, thousands, millions or billions of years (as the case) they reunite again. The relationship between A and B could start as parasitism, predation, proto cooperation, mastery of one people by another, beginning of commercial relationships, etc. I argue that if sufficient amount of time is allowed to pass, most relationships will tend to symbiosis as cooperation will be beneficial for both groups—but indeed complexity can arise from any other type of relationship. Predators, for instance, are beneficial for their prey as they hunt, preferably, weak individuals keeping the pray group healthy (Genovart et al., 2010). In the other hand, predators could drive prey communities—and vice versa—to complexity via the first mechanism mentioned above, in which group selection has a leading role (Fryxell et al., 2007).

Even parasitic relationships may also turn beneficial with time, as simplification and interdependence gradually install. For instance, the hygiene hypothesis states that modern day absence of exposure of humans to intestinal helminthes (worms) due to excessive hygiene appears to be an important environmental factor contributing to development of various illnesses, such as Inflammatory Bowel Disease, Bronchial Asthma and autoimmune diseases (Okada et al., 2010; Jouvin and Kinet, 2012; Zaccone and Cooke, 2013). The parasite produces substances that regulate the immune system's circuitry in order to dampen its effects over the worm's environment, resulting in benefit to the host. In turn, the host (us) has become simpler. Now we depend on parasites to modulate our own immune systems and maintain healthy! What an apparent paradox!

Irrespective of the kind of relation, with time, fusion of A and B forms a single thing again, but this time much more complex. Further specialization, progressive simplification and subsequent increment of interdependence within the group will continue to take place in most cases. Examples of this mechanism are the emergence of the eukaryotic cell by endosymbiosis, symbiosis of humans — and other animals — with intestinal microbiota (Mathis and Benoist, 2011), and the economic and social success of the United States of America, an admixture of diverse groups of immigrants. Ecosystems, in turn, are complex systems composed of innumerable subsystems within which we can see both kinds of sequences.

Within the context of the theory exposed here, the expressions "group selection" or "individual selection" are relative. Individual selection is also group selection relative to the nearest lower level of complexity. At any level, group performance may be enhanced from the creation of complexity by these mechanisms as the division of tasks is energy efficient. Further specialization and simplification should refine diverse strategies for survival of the group, such as the improvement of communication methods between elements, and the emergence of control (tonic, feedback or antagonistic) mechanisms that provide stability and resilience to the system. Thus, in the study of evolution toward complexity, the evolution of the individuals' environment determined by the enhancement of "social" conditions might be as important as the evolution of traits at the individual level, as both are intimately interwoven.

The skier falling down the hill

To illustrate in a different way how the formation of complexity is coupled to energy and entropy, I will take the risk of making a generalization of a lesson in elementary physics we learn at high school. The aforementioned lesson is about potential energy and kinetic energy. Kinetic energy — the teacher told us — is the one possessed by a moving object, while potential energy is stored energy. As a skier climbs up the hill on the cable car he stores potential energy. Upon reaching the top, potential energy is at its maximum. Once he slides down the slope begins to gain kinetic energy, and at the same time potential energy decreases. The paths that the skier can take are several but he will take just one. The skier cannot turn around and go up the hill as it falls, except for a very small distance after which he begin to fall again. Such temporal ascent represents activation energy in chemical reactions or energy and intelligence investment in complexity. At the lowest point, the total energy — kinetic plus potential — is minimal and all will be dissipated as heat (entropy).

At this point, the reader is no stranger to the fact that all living beings are essentially composed of proteins, humans included. It is thought that there are over 100,000 different proteins comprising the human genome (however the true number is still to be elucidated; Barabási et al., 2011) which are those that perform multiple tasks for survival and adequate functioning of our organism. A protein is a chain of aminoacids. There are a total of 20 types of amino acids and the sequential combination of these in the chain is what (basically) differentiates a protein from another. A gene is a plan or scheme that will determine the order of amino acids in the protein in the same way that blueprints of a house tell us the position of every wall, window or door. The ability of an aminoacid to form bonds with others resulting in linear-chain constructions should have been one of the earliest manifestations of life on Earth, as discussed in a previous section.

However it is not the order or sequence of amino acids within the chain what gives a specific protein its function, but an emergent property of a higher level of complexity. The protein can carry out its work within the body due to its specific form, i.e., its three-dimensional projection. It is the shape (together with superficial forces) what allows it to bind to other proteins and chemical substrates or to form complex structures inside and outside cells (Halabi et al., 2009). But how does a specific protein acquires its structure?

The spatial projection of the protein depends directly on the aminoacid sequence in the chain. Each of the 20 types of aminoacids has particular characteristics and properties. There are negatively as well as positively charged aminoacids; some are hydrophobic and other hydrophilic (with low and high affinity for water, respectively). A protein molecule is built inside the cell in a linear fashion, but this is an unstable configuration with a lot of internal (potential) energy which is equivalent to the skier at the highest point of the hill. Then, in an instant, the molecule "falls down the slope". The hydrophobic amino acids rejected by water molecules hide in the center while hydrophilic ones rapidly float to the surface. The positively charged aminoacids approach and relate to the negatively charged; the ones with same charge repel each other, and while all of these things happen the molecule performs innumerable febrile contortions until, some milliseconds later, gains its final three-dimensional shape with minimal internal energy (Gebhardt et al., 2010). This whole process, called protein folding, is mostly spontaneous; however, the correct folding is guided by a set of "physician" proteins called chaperones (Hartl and Hayer-Hartl, 2009), which is a great example of occupational complexity at the molecular level.

As the skier can take several paths on his way down the slope, a specific protein can be folded in several possible ways, namely "energy landscapes" (Wolynes et al., 2012). Some of these forms are functional, and some are not. Knowledge of the functional shapes of proteins is of capital importance for designing new drugs, and its determination given the aminoacid sequence one of the most arduous tasks of molecular biologists and bioinformatics for which a lot of time and computational power is necessary (Kelley and Sternberg, 2009). Fortunately, with the advent of internet it is possible for the public to voluntarily yield computational time of PCs for these and other scientific works. Like thousands of ants in the anthill, computers act jointly to clarify the final configuration of a protein (voluntary donation of PC time can be done at http://boinc.berkeley.edu/). Moreover, recently researchers have designed games and puzzles which are downloaded to the network so the public can play their free time assembling proteins (at http://fold.it/portal). It turns out that humans are still better than computers in this work! It is envisaged that in the near future the solution to the biggest problems faced by humanity shall be solved by way of games distributed in the network.

Not only proteins begin their lives in the highest point of the hill; humans too. Shortly after the sperm fertilizes the egg and both pronuclei fuse to form a zygote, cell division begins. At first a two-celled embryo is formed, then four-celled, eight, sixteen, etc. Each of these cells during the initial embryonic stages is totipotent, i.e. capable by itself to form a complete embryo. In a later time of embryonic development, the cells turn to be no longer totipotent but still capable of forming all types of tissues derived from the three embryonic layers (pluripotency). Then, further differentiation makes cells capable of forming just specific types of tissue (multipotency) (Mitalipov and Wolf, 2009; Zhang and Kilian, 2013). At later stages additional differentiation originates specific organs and body parts. As each cell gets specialized, acquiring functions that enable it to be part of a particular tissue, the embryo falls down the slope with no possibility of climbing back by its own. During fetal stage, after birth and until the end of our lives we continue falling, with a population of stem cells replenishing every tissue as differentiated cells die. Today, we have learned to reverse the process in the laboratory by means of stem cells technology. It is now possible to reverse a differentiated, specialized cell into a pluripotent cell, capable of deriving into any tissue using combinations of transcription factors. Regarding developmental biology and regenerative medicine, we are learning to turn back and climb the hill for a moment (Adachi and Schöler, 2013; Ben-David et al., 2013).

Human (or any other multicelular species) development from one cell is a great example of simplification within a level to form complexity in the next upper level. Individual cells originate from a complex (totipotent) zygote. Then, gradually specialize silencing the rest of their functions (simplification). Meanwhile, the embryo as a whole gains complexity. No other process on earth shows in its entire splendor the harmony between entropy and natural forces acting in conjunction to form complexity.

It is thought that he universe itself began as a singularity of enormous potential (negative) energy of a magnitude equal to the total sum of its rest energy (mass) and kinetic energy, originated from the quantum fluctuations of the vacuum (Berman and Trevisan, 2010); so, it has no other way than to fall irreversibly down the hill, like the skier. However, for some reason that still seems supremely mysterious to me, while falling beautiful forms of increasing complexity are created.

Human relationships

A diversity of types of relations exists among humans. Some are superfluous and transient, like commercial and other business relations; but others, like the ones with work or study mates are more profound. Relations can be horizontal, as with brothers, sisters, cousins and peers; or they can be vertical as those between bosses and subordinates, or parents and children. Friends are the family that we wanted to have, but in general, the strongest ties are within the family, generally lasting a lifetime. Powerful emotional attachments of various kinds bring us close to other humans. A fundamental attribute of humans is our enhanced ability to form extended family groups with numerous members including in-laws and friends, for which we deserve a position among eusocial species (Hardisty and Cassill, 2010; Wilson, 2012).

The relatively new telecommunication network has allowed us to approach each other even more and increase the number of interpersonal relationships. Friends in a social networking site (e.g. Facebook) are numbered in the hundreds, even thousands, increasing our range of social action, the number of dyadic relationships and the complexity of society (Viswanath et al, 2009; Omoush et al., 2012). These relationships have the weakness of being more relaxed and less profound than those "face to face", but no matter if our assessment of that trend is positive or negative, its reversal is unlikely.

One cannot downplay the importance of the quality of human relationships. Just as the chemical bonds between elements and the production of signaling substances between cells aim to create complexity at the molecular and organismal levels respectively, human relations constitute the main amalgamating element of human complex societies. Cognitive features that allowed us living in complex and efficient groups were improved early in evolution of *Homo sapiens,* as the development of structures in human (in relation to ape) brains, such as an area specialized in face recognition located in the occipito-temporal cortex and other cerebral regions, such as hippocampi and the right inferior frontal area (Taylor et al., 2011); and mirror neurons involved in learning by imitation as well as in empathetic behavior (Iacoboni, 2009; Patel, 2011).

Wilson (2012) cites Sage Rabbi Hillel, the renowned Jewish scholar, when someone once challenged him to explain the Torah standing on one foot. The wise man did not decline the challenge, and while balancing on one foot said: "That which is hateful to you, do not do to your fellow. That is the whole Torah; the rest is the explanation; go and learn". It is no coincidence that the Bible also tells us that when Jesus was asked by a Pharisee about the greatest commandment of the law, he responded: "Love the Lord your God with all your heart and with all your soul and with all your mind and with all your strength." But then he claimed: "The second is this: 'Love your neighbor as yourself.' There is no commandment greater than these."

The "Golden Rule", the ultimate expression of empathy, is repeated in all major religions including Buddhism, Confucianism, Zoroastrianism, Judaism, Christianity, Islam and Taoism. And this need for empathy and good relations was not only recognized as a key element in human life by Jesus Christ and the great spiritual teachers, but also by philosophers of all time.

When British philosopher Bertrand Russell was asked in a television interview about what were the things he considered future generations should know from the life he lived and the lessons he had learned from it, he responded:

"I should like to say two things: one intellectual and one moral... The moral thing I should wish to say, I should say love is wise, hatred is foolish. In this world which is getting more closely and closely interconnected we have to learn to tolerate each other; we have to learn to put up with the fact that some people say things that we don't like. We can only live together in that way and if we are to live together and not die together we must learn a kind of charity and a kind of tolerance which is absolutely vital to the continuation of human life on this planet."

Few would have said it better and with such authority as Bertrand Russell.

Personally, I think that relations with our neighbors are infinitesimal bridges that make us an integral part of the great river of life on its course towards higher levels of complexity; longitudinal bridges that form networks, and vertical bridges linking past generations to future ones. Relationships are the very fabric of which life is built. And today, good relations have become absolutely necessary for our survival as a species, and in addition, to take the next big leap: the development of the intelligent Global Superorganism.

LITERATURE CITED

Abbot P, Abe J, Alcock J, Alizon S, Alpedrinha JA et al. (2011).

Inclusive fitness theory and eusociality. Nature. 471(7339):E1-E4.

Adachi K, Schöler HR (2012). Directing reprogramming to pluripotency by transcription factors. Current Opinion in Genetics & Development. En Imprenta.

American Psychiatric Association (1995). DSM-IV. Manual diagnóstico y estadístico de los trastornos mentales. Barcelona: Masson.

Arendt D (2008). The evolution of cell types in animals: emerging principles from molecular studies. Nature Reviews Genetics. 9(11):868-882.

Asimov I (1956). The Last Question. Science Fiction Quarterly. Noviembre.

Bak P, Tang C, Wiesenfeld K (1987). Self-organized criticality: An explanation of the 1/f noise. Physical review letters. 59(4):381-384.

Barabási AL, Gulbahce N, Loscalzo J (2011). Network medicine: a network-based approach to human disease. Nature Reviews Genetics. 12(1):56-68.

Bashe CJ, Johnson LR, Palmer JH, Pugh EW (1986). IBM's early computers. MIT press.

Beauchamp C (2010). Who Invented the Telephone? : Lawyers,

Patents, and the Judgments of History. Technology and Culture. 51(4):854-878.

Beekman M, Blanché H, Perola M, Hervonen A, Bezrukov V et al. (2013). Genome-wide linkage analysis for human longevity: Genetics of Healthy Aging Study. Aging cell. 12(2):184-93

Ben-David U, Nissenbaum J, Benvenisty N (2013). New Balance in Pluripotency: Reprogramming with Lineage Specifiers. Cell. 153(5):939-940.

Berkhout J, Bosdriesz E, Nikerel E, Molenaar D, de Ridder D et al. (2013). How biochemical constraints of cellular growth shape evolutionary adaptations in metabolism. Genetics. 194(2):505-512.

Berman MS, Trevisan LA (2010). On the Creation of Universe out of Nothing. International Journal of Modern Physics D. 19(08n10):1309-1313.

Boole G (1847). The mathematical analysis of logic. Being an essay toward a calculus of deductive reasoning. Philosophical Library.

Brunk GG (2002). Why do societies collapse? A theory based on self-organized criticality. Journal of Theoretical Politics. 14(2):195-230.

Byers PH (2002). Killing the messenger: new insights into nonsense-mediated mRNA decay. Journal of Clinical

Investigation. 109(1):3-6.

Campbell-Kelly M, Aspray W, Wilkes MV (2004). Computer: a history of the information machine (Vol. 2). Boulder: Westview Press.

Cavalli-Sforza LL (1997). Genes, peoples, and languages. Proceedings of the National Academy of Sciences USA. 94(15):7719-7724.

Cavalli-Sforza LL, Piazza A, Menozzi P, Mountain J (1988). Reconstruction of human evolution: bringing together genetic, archaeological, and linguistic data. Proceedings of the National Academy of Sciences USA. 85(16):6002-6006.

Chamary JV, Parmley JL, Hurst LD (2006). Hearing silence: non-neutral evolution at synonymous sites in mammals. Nature Reviews Genetics. 7(2):98-108.

Chapman C, Laird J, KewalRamani A (2013). Trends in high school dropout and completion rates in the United States: 1972-2009. BiblioGov.

Chu S, DeRisi J, Eisen M, Mulholland J, Botstein D et al. (1998). The transcriptional program of sporulation in budding yeast. Science. 282(5389):699-705.

Ciechanover A (2005). Proteolysis: from the lysosome to ubiquitin and the proteasome. Nature Reviews Molecular Cell Biology. 6(1):79-87.

Clarke TC, Bolton S J (2010). The planets and our culture a history and a legacy. Proceedings of the International Astronomical Union. 6(S269): 199-212.

Colantonio S, Lasker GW, Kaplan BA, Fuster V (2003). Use of surname models in human population biology: a review of recent developments. Human Biology. 75(6):785-807.

Cole LK, Ross LS (2001). Apoptosis in the developing zebrafish embryo. Developmental biology. 240(1):123-142.

Collins FS, Lander ES, Rogers J, Waterston RH, International Genome Sequencing Consortium. (2004). Finishing the euchromatic sequence of the human genome. Nature. 431(7011):931-945.

Comas D, Calafell F, Mateu E, Pérez-Lezaun A, Bosch E, et al. (1998). Trading genes along the silk road: mtDNA sequences and the origin of central Asian populations. The American Journal of Human Genetics. 63(6):1824-1838.

Cooper MD, Alder MN (2006). The evolution of adaptive immune systems. Cell. 124(4):815-822.

Crow JF, Mange AP (1965). Measurement of inbreeding from the frequency of marriages between persons of the same surname. Biodemography and Social Biology. 12(4):199-203.

Dafallah SE, Ambago J, El-Agib F (2003). Obstructed labor in a teaching hospital in Sudan. Saudi medical journal.

24(10):1102-1104.

Darwin C (1859). On the Origin of Species.

De Navarro JM (1925). Prehistoric routes between northern Europe and Italy defined by the amber trade. The Geographical Journal. 66(6):481-503.

Dediu D (2013). Genes: Interactions with Language on Three Levels—Inter-Individual Variation, Historical Correlations and Genetic Biasing. in: The Language Phenomenon. Springer Berlin Heidelberg.

Deeg M. (1999). Multiculturalism in Asian religions: North India. Central Asia and China in ancient times. Ingår i Diskus. 5.

Driver L (2010). What Made Us Human: Analysis of Richard Wrangham's Cooking Hypothesis. Lambda Alpha Journal. 40:21.

Dudley SS (2010). Drug trafficking organizations in Central America: transportistas, Mexican cartels and maras. Shared Responsibility. 9.

Fish JL, Lockwood CA (2003). Dietary constraints on encephalization in primates. American journal of physical anthropology. 120(2):171-181.

Fitch WM (1970). Distinguishing homologous from analogous proteins. Systematic Zoology. 19:99–106.

Foner N (2000). From Ellis Island to JFK: New York's two great waves of immigration. Yale University Press.

Foster K, Shaulsky G, Strassmann JE, Queller DC, ThompsonCRL (2004). Pleiotropy as a mechanism to stabilize cooperation. Nature. 431:693-696.

Frazer KA, Elnitski L, Church DM, Dubchak I, Hardison RC (2003). Cross-species sequence comparisons: a review of methods and available resources. Genome Research. 13(1):1-12.

Fryxell JM, Mosser A, Sinclair AR, Packer C (2007). Group formation stabilizes predator–prey dynamics. Nature. 449(7165):1041-1043.

Gabaldon T, Koonin EV (2013). Functional and evolutionary implications of gene orthology. Nature Reviews Genetics. 14(5):360-366.

Gaym A (2002). Obstructed labor at a district hospital. Ethiopian medical journal. 40(1):11.

Ge H, Qian H (2011). Heat Dissipation and Self-consistent Nonequilibrium Thermodynamics of Open Driven Systems. arXiv preprint arXiv:1106.2564.

Gebhardt J CM, Bornschlögl T, Rief M (2010). Full distance-resolved folding energy landscape of one single protein molecule. Proceedings of the National Academy of Sciences

USA. 107(5):2013-2018.

Genovart M, Negre N, Tavecchia G, Bistuer A, Parpal L. et al. (2010). The young, the weak and the sick: evidence of natural selection by predation. PloS one. 5(3):e9774.

Germonpré M, Sablin M, Khlopachev GA, Grigorieva GV (2008). Possible evidence of mammoth hunting during the Epigravettian at Yudinovo, Russian Plain. Journal of Anthropological Archaeology. 27(4):475-492.

Gibson A (2011). A new view of the birth of Homo sapiens. Science. 331(6016):392-394.

Gilad Y, Man O, Glusman G (2005). A comparison of the human and chimpanzee olfactory receptor gene repertoires. Genome research. 15(2):224-230.

Gilbert SF, Opitz JM, Rudolf AR (1996). Resynthesizing Evolutionary and Developmental Biology. Developmental biology. 173:357–372

Gjørret JO, Knijn HM, Dieleman SJ, Avery B, Larsson LI et al. (2003). Chronology of apoptosis in bovine embryos produced in vivo and in vitro. Biology of reproduction. 69(4):1193-1200.

Gleiser PM, Spoormaker VI (2010). Modelling hierarchical structure in functional brain networks. Philosophical Transactions of the Royal Society A: Mathematical, Physical and Engineering Sciences. 368(1933):5633-5644.

Glickman MH, Ciechanover A (2002). The ubiquitin-proteasome proteolytic pathway: destruction for the sake of construction. Physiological reviews. 82(2): 373-428.

Gommans WM, Mullen SP, Maas S (2009). RNA editing: a driving force for adaptive evolution? Bioessays. 31(10):1137-1145.

Goodenough WH (1997). Phylogenetically related cultural traditions. Cross-Cultural Research. 31(1):16-26.

Grabowski MW (2013). Hominin obstetrics and the evolution of constraints. Evolutionary Biology. 40(1):57-75.

Grizzi F, Colombo P, Taverna G, Chiriva-Internati M, Cobos E et al. (2007). Geometry of human vascular system: is it an obstacle for quantifying antiangiogenic therapies? Applied Immunohistochemistry & Molecular Morphology, 15(2):134-139.

Guimaraes-Souza NK, Yamaleyeva LM, AbouShwareb T, Atala A, Yoo JJ (2012). In vitro reconstitution of human kidney structures for renal cell therapy. Nephrology Dialysis Transplantation. 27(8):3082-3090.

Gutierrez H, Davies AM (2011). Regulation of neural process growth, elaboration and structural plasticity by NF-κB. Trends in neurosciences. 34(6):316-325.

Hadany L, Comeron JM (2008). Why are sex and recombination so common? Annals of New York Academy of Sciences. 1133:26-43.

Halabi N, Rivoire O, Leibler S, Ranganathan R (2009). Protein sectors: evolutionary units of three-dimensional structure. Cell. 138(4):774-786.

Handley LJL, Manica A, Goudet J, Balloux F (2007). Going the distance: human population genetics in a clinal world. Trends in Genetics. 23(9):432-439.

Hardisty BE, Cassill DL (2010). Extending eusociality to include vertebrate family units. Biology and Philosophy. 25(3):437-440.

Hartl FU, Hayer-Hartl M (2009). Converging concepts of protein folding in vitro and in vivo. Nature structural & molecular biology. 16(6):574-581.

Hendler J, Berners-Lee T (2010). From the Semantic Web to social machines: A research challenge for AI on the World Wide Web. Artificial Intelligence. 174(2):156-161.

Herrera-Paz EF, Matamoros M, Carracedo Á (2010). The Garífuna (Black Carib) people of the Atlantic coasts of Honduras: Population dynamics, structure, and phylogenetic relations inferred from genetic data, migration matrices, and isonymy. American Journal of Human Biology. 22(1):36-44.

Herrera-Paz EF (2013). Apellidos e isonimia en las comunidades garífunas de la costa atlántica de Honduras. Revista Médica del Instituto Mexicano del Seguro Social. 51(2):150-7.

Higginson DM, Pitnick S (2011). Evolution of intra-ejaculate sperm interactions: do sperm cooperate? Biological Reviews. 86(1):249-270.

Hill RS, Walsh CA (2005). Molecular insights into human brain evolution. Nature. 437(7055):64-67.

Hirata S, Fuwa K, Sugama K, Kusunoki K, Takeshita H (2011). Mechanism of birth in chimpanzees: humans are not unique among primates. Biology letters. 7(5):686-688.

Hirth KG (1978). Interregional trade and the formation of prehistoric gateway communities. American Antiquity. 35-45.

Hochfelder D (2010). Two controversies in the early history of the telegraph. Communications Magazine, IEEE. 48(2):28-32.

Holling CS (1973). Resilience and stability of ecological systems. Annual review of ecology and systematics. 4:1-23.

Horvath P, Barrangou R (2010). CRISPR/Cas, the immune system of bacteria and archaea. Science. 327(5962):167-170.

Huising MO, Kruiswijk CP, Flik G (2006). Phylogeny and evolution of class-I helical cytokines. Journal of

Endocrinology. 189(1):1-25.

Hunt J, Breuker CJ, Sadowsk JA, Moore AJ (2009). Male–male competition, female mate choice and their interaction: determining total sexual selection. Journal of evolutionary biology. 22(1):13-26.

Iacoboni M (2009). Imitation, empathy, and mirror neurons. Annual review of psychology. 60:653-670.

Imbimbo A (2009). Steve Jobs: The Brilliant Mind Behind Apple. Gareth Stevens Publishing.

Jacobi FK, Pusch CM (2010). A decade in search of myopia genes. Frontiers in bioscience: a journal and virtual library. 15:359.

Jakobsson Á (2008). A contest of cosmic fathers. Neophilologus. 92(2):263-277.

Jeknic-Dugic J, Dugic M, Francom A, Arsenijevic M (2012). Quantum Structures of the Hydrogen Atom. arXiv preprint arXiv :1204.3172.

Jobling MA (2001). In the name of the father: surnames and genetics. TRENDS in Genetics. 17(6):353-357.

Jouvin MH, Kinet JP (2012). Trichuris suis ova: Testing a helminth-based therapy as an extension of the hygiene hypothesis. Journal of Allergy and Clinical Immunology. 130(1):3-10.

Kaplan DR, Miller FD (2000). Neurotrophin signal transduction in the nervous system. Current opinion in neurobiology. 10(3):381-391.

Kelley LA, Sternberg MJ (2009). Protein structure prediction on the Web: a case study using the Phyre server. Nature protocols. 4(3):363-371.

Kennedy SR, Loeb LA, Herr AJ (2012). Somatic mutations in aging, cancer and neurodegeneration. Mechanisms of ageing and development. 133(4):118-126.

Kohnert D (2007). African migration to Europe: obscured responsibilities and common misconceptions. GIGA Working Papers.

Kohonen-Corish MR, Al-Aama JY, Auerbach AD, Axton M, Barash CI et al. (2010). How to catch all those mutations—the report of the Third Human Variome Project Meeting, UNESCO Paris, May 2010. Human mutation. 31(12):1374-1381.

Koonin EV (2005). Orthologs, paralogs, and evolutionary genomics 1. Annual Review Genetics. 39: 309-338.

Kumar H, Kawai T, Akira S (2011). Pathogen recognition by the innate immune system. International Reviews of Immunology. 30(1):16-34.

Kurzweil R. (2003). Exponential growth an illusion. Response to

Ilkka Tuomi essay (September 23) http://www. kurzweilai. net/meme/frame.html.

Kutschera U, Niklas KJ (2004). The modern theory of biological evolution: an expanded synthesis. Naturwissenschaften. 91(6):255-276.

Ladd TD, Jelezko F, Laflamme R, Nakamura Y, Monroe C, O'Brien JL (2010). Quantum computers. Nature. 464(7285):45-53.

Lander ES (2011). Initial impact of the sequencing of the human genome. Nature. 470(7333):187-197.

Lathrap DW (1973). The antiquity and importance of long-distance trade relationships in the moist tropics of pre-Columbian South America. World Archaeology. 5(2):170-186.

Leiner BM, Cerf VG, Clark DD, Kahn RE, Kleinrock L et al. (2009). A brief history of the Internet. ACM SIGCOMM Computer Communication Review. 39(5):22-31.

Levy S, Sutton G, Ng PC, Feuk L, Halpern AL et al. (2007). The diploid genome sequence of an individual human. PLoS biology. 5(10):e254.

Li JZ, Absher DM, Tang H, Southwick AM, Casto AM et al. (2008). Worldwide human relationships inferred from genome-wide patterns of variation. Science. 319(5866):1100-1104.

Li Y, Vinckenbosch N, Tian G, Huerta-Sanchez E, Jiang T et al. (2010). Resequencing of 200 human exomes identifies an excess of low-frequency non-synonymous coding variants. Nature genetics. 42(11):969-972.

Lin H, Shuai JW (2010). A stochastic spatial model of HIV dynamics with an asymmetric battle between the virus and the immune system. New Journal of Physics. 12(4):043051.

Liu X (2001). The Silk Road: overland trade and cultural interactions in Eurasia. in: Agricultural and Pastoral Societies in Ancient and Classical History. 151-79. Philadelphia. Temple University Press.

Longo VD, Mitteldorf J, Skulachev VP 2005. Programmed and altruistic ageing. Nature Reviews Genetics. 6:866–872.

Lopes R, Betrouni N (2009). Fractal and multifractal analysis: A review. Medical image analysis. 13(4):634.

MacArthur DG, Balasubramanian S, Frankish A, Huang N, Morris J et al. (2012). A systematic survey of loss-of-function variants in human protein-coding genes. Science. 335(6070):823-828.

Magon N, Kalra S (2011). The orgasmic history of oxytocin: Love, lust, and labor. Indian Journal of Endocrinology and Metabolism. 15(Suppl3):S156.

Mandelbrot BB (1982). The Fractal Geometry of Nature. New

York: WH Freeman and Co.

Marian AJ (2013). Errors in DNA replication and genetic diseases. Current opinion in cardiology. 28(3):269-271.

Marks SJ, Levy H, Martinez-Cadenas C, Montinaro F, Capelli C (2012). Migration distance rather than migration rate explains genetic diversity in human patrilocal groups. Molecular Ecology. 21(20):4958-4969.

Massey DS (2002). A brief history of human society: The origin and role of emotion in social life. American Sociological Review. 67(1):1-29.

Mathis D, Benoist C (2011). Microbiota and autoimmune disease: the hosted self. Cell Host & Microbe. 10(4):297-301.

Mckitterick R. (2000). Books and sciences before print. in: Books and the sciences in history.

McNally JG, Mazza D (2010). Fractal geometry in the nucleus. The EMBO journal. 29(1):2.

Mesoudi A, Whiten A, Laland KN (2004). Perspective: is human cultural evolution Darwinian? Evidence reviewed from the perspective of The Origin of Species. Evolution. 58(1):1-11.

Mesoudi A, Whiten A, Laland KN (2006). Towards a unified science of cultural evolution. Behavioral and Brain Sciences. 29(4):329-346.

Mitalipov S, Wolf D (2009). Totipotency, pluripotency and nuclear reprogramming. in:Engineering of Stem Cells (pp. 185-199). Springer Berlin Heidelberg.

Moazed D (2009). Small RNAs in transcriptional gene silencing and genome defense. Nature. 457(7228):413-420.

Moore GE (1998). Cramming more components onto integrated circuits. Proceedings of the IEEE. 86(1):82-85.

Nicolis G, Prigogine I (1971). Fluctuations in nonequilibrium systems. Proceedings of the National Academy of Sciences USA. 68(9):2102-2107.

Nomiyama H, Osada N, Yoshie O (2010). The evolution of mammalian chemokine genes. Cytokine & Growth Factor Reviews. 21(4):253-262.

Norddahl GL, Pronk CJ, Wahlestedt M, Sten G, Nygren JM et al. (2011). Accumulating mitochondrial DNA mutations drive premature hematopoietic aging phenotypes distinct from physiological stem cell aging. Cell stem cell. 8(5):499-510.

Nowak MA, Tarnita CE, Wilson EO (2010). The evolution of eusociality. Nature. 466(7310):1057-1062.

Okada H, Kuhn C, Feillet H, Bach JF (2010). The 'hygiene hypothesis' for autoimmune and allergic diseases: an update. Clinical & Experimental Immunology. 160(1):1-9.

Omoush A, Saleh K, Yaseen SG, Atwah Alma'aitah M (2012). The impact of Arab cultural values on online social networking: The case of Facebook. Computers in Human Behavior. 28(6):2387-2399.

Orr HA (2010). The population genetics of beneficial mutations. Philosophical Transactions of the Royal Society B: Biological Sciences. 365(1544):1195-1201.

Oppenheimer S (2012). Out-of-Africa, the peopling of continents and islands: tracing uniparental gene trees across the map. Philosophical Transactions of the Royal Society B: Biological Sciences, 367(1590):770-784.

Park IH, Zhao R, West JA, Yabuuchi A, Huo H et al. (2007). Reprogramming of human somatic cells to pluripotency with defined factors. Nature. 451(7175):141-146.

Parpola S (2004). National and ethnic identity in the Neo-Assyrian empire and Assyrian identity in post-empire times. Journal of Assyrian Academic Studies. 18(2):5-22.

Pascual V, Chaussabel D, Banchereau J (2010). A genomic approach to human autoimmune diseases. Annual review of immunology. 28:535.

Patel T (2011). Mirror Neurons: Recognition, Interaction, Understanding. Berkeley Scientific Journal. 14(2).

Pearson H (2007). Meet the human metabolome. Nature.

446(7131):8.

Peck JR (1994). A ruby in the rubbish: beneficial mutations, deleterious mutations and the evolution of sex. Genetics. 137(2):597-606.

Perry DA (1995). Self-organizing systems across scales. Trends in Ecology & Evolution. 10(6):241-244.

Pestka S, Krause CD, Sarkar D, Walter MR, Shi Y et al. (2004). Interleukin-10 and related cytokines and receptors. Annual Reviews of Immunology. 22:929-979.

Pfaus JG, Kippin TE, Centeno S (2001). Conditioning and sexual behavior: a review. Hormones and Behavior. 40(2):291-321.

Phillips PK, Heath JE (1995). Dependency of surface temperature regulation on body size in terrestrial mammals. Journal of Thermal Biology. 20(3):281-289.

Pierce SK, Miller LH (2009). World Malaria Day 2009: what malaria knows about the immune system that immunologists still do not. The Journal of Immunology. 182(9):5171-5177.

Prigogine I, Lefever R (1968). Symmetry breaking instabilities in dissipative systems. II. The Journal of Chemical Physics. 48:1695.

Rachlin H (2002). Altruism and selfishness. Behavioral and Brain

Sciences. 25(2):239-250.

Raup DM (1986). Biological extinction in earth history. Science. 231(4745):1528-1533.

Raynal NJM, Si J, Taby RF, Gharibyan V, Ahmed S et al. (2012). DNA methylation does not stably lock gene expression but instead serves as a molecular mark for gene silencing memory. Cancer research. 72(5):1170-1181.

Read LE (1958). I, the pencil. NY: Irvington-on-Hudson.

Rock WP, Sabieha AM, Evans RIW (2006). A cephalometric comparison of skulls from the fourteenth, sixteenth and twentieth centuries. British dental journal. 200(1):33-37.

Rodgers D, Muggah R, Stevenson C (2009). Gangs of Central America: causes, costs, and interventions. Small Arms Survey. Graduate Institute of International and Development Studies. Geneva.

Rosenberg K, Trevathan W (2002). Birth, obstetrics and human evolution. BJOG: An International Journal of Obstetrics & Gynaecology. 109(11):1199-1206.

Roth G, Dicke U (2005). Evolution of the brain and intelligence. Trends in cognitive sciences. 9(5):250-257.

Sastry PS, Rao KS (2000). Apoptosis and the nervous system. Journal of neurochemistry. 74(1):1-20.

Saw SM, Chua WH, Wu HM, Yap E, Chia KS et al. (2000). Myopia: gene-environment interaction. Annals of the Academy of Medicine. Singapore. 29(3):290.

Scapoli C, Mamolini E, Carrieri A, Rodriguez-Larralde A, Barrai I (2007). Surnames in Western Europe: A comparison of the subcontinental populations through isonymy. Theoretical population biology. 71(1):37-48.

Schiller NG, Basch L, Blanc-Szanton C (1992). Transnationalism: A new analytic framework for understanding migration. Annals of the New York Academy of Sciences. 645(1):1-24.

Schneider ED, Kay JJ (1994). Life as a manifestation of the second law of thermodynamics. Mathematical and computer modelling. 19(6):25-48.

Schneider ED, Kay JJ (1995). Order from disorder: the thermodynamics of complexity in biology. in: What is life? The next fifty years: Speculations on the future of biology. 161-172.

Schwartz SH, Huismans S (1995). Value priorities and religiosity in four western religions. Social Psychology Quterly. 58(2):88-107.

Shumway W (1932). The recapitulation theory. The Quarterly Review of Biology. 7(1):93-99.

Shyu AB, Wilkinson MF, van Hoof A (2008). Messenger RNA

regulation: to translate or to degrade. The EMBO journal. 27(3):471-481.

Small RL, Cronn RC, Wendel JF (2004). LAS Johnson Review No. 2. Use of nuclear genes for phylogeny reconstruction in plants. Australian Systematic Botany. 17(2):145-170.

Smith BD, Zeder MA (2013). Anthropocene. The Onset of the Anthropocene. En Imprenta.

Smith TW (1990). Classifying protestant denominations. Review of Religious Research. 31(3):225-245.

Sober E, Wilson DS (2011). Adaptation and natural selection revisited. Journal of Evolutionary Biology. 24(2):462-468.

Sole RV, Manrubia SC, Benton M, Bak P (1997). Self-similarity of extinction statistics in the fossil record. Nature. 388(6644):764-767.

Sorci G, Cornet S, Faivre B (2013). Immune Evasion, Immunopathology and the Regulation of the Immune System. Pathogens. 2(1):71-91.

Spencer J, Thomas MSC, McClelland JL (2009). Toward a unified theory of development: connectionism and dynamic systems theory re-considered. Oxford/Nueva York: Oxford University Press.

Spielman D, Brook BW, Frankham R (2004). Most species are not driven to extinction before genetic factors impact them. Proceedings of the National Academy of Sciences USA. 101(42):15261-15264.

Stern A, Sorek R (2011). The phage-host arms race: Shaping the evolution of microbes. Bioessays. 33(1): 43-51.

Stewart C (1999). Syncretism and its synonyms: Reflections on cultural mixture. Diacritics. 29(3): 40-62.

Stower H (2012). Chromosome biology: Pairing up for the genetic exchange. Nature Reviews Genetics. 13(7):449-449.

Swade D, Babbage C (2001). Difference Engine: Charles Babbage and the Quest to Build the First Computer. Viking Penguin.

Taylor MJ, Mills T, Pang EW (2011). The development of face recognition; hippocampal and frontal lobe contributions determined with MEG. Brain topography. 24(3-4):261-270.

Temin P (2001). A market economy in the early Roman Empire. The Journal of Roman Studies. 91:169-181.

Tetel MJ, Pfaff DW (2010). Contributions of estrogen receptor-α and estrogen receptor-β to the regulation of behavior. Biochimica et Biophysica Acta (BBA)-General Subjects. 1800(10):1084-1089.

Tijsterman M, Ketting RF, Plasterk RH (2002). The genetics of RNA silencing. Annual Review of Genetics. 36(1):489-519.

Todman D (2007). A history of caesarean section: from ancient world to the modern era. Australian and New Zealand Journal of Obstetrics and Gynaecology. 47(5):357-361.

Vaissière T, Sawan C, Herceg Z (2008). Epigenetic interplay between histone modifications and DNA methylation in gene silencing. Mutation Research/Reviews in Mutation Research. 659(1):40-48.

Vespignani A (2010). Complex networks: The fragility of interdependency. Nature. 464(7291):984-985.

Vidal M, Cusick ME, Barabasi AL (2011). Interactome networks and human disease. Cell. 144(6):986-998.

Vijg J, Suh Y (2013). Genome instability and aging. Annual review of physiology. 75(1).

Viswanath B, Mislove A, Cha M, Gummadi KP (2009). On the evolution of user interaction in facebook. in: Proceedings of the 2nd ACM workshop on Online social networks (pp. 37-42). ACM.

Von Cramon-Taubadel N, Lycett SJ (2008). Brief communication: human cranial variation fits iterative founder effect model with African origin. American Journal of Physical Anthropology. 136(1):108-113.

Wagner A (2011). Genotype networks shed light on evolutionary constraints. Trends in ecology & evolution. 26(11):577-584.

Wagner CS, Leydesdorff L (2005). Network structure, self-organization, and the growth of international collaboration in science. Research policy. 34(10):1608-1618.

Wells C (1975). Ancient obstetric hazards and female mortality. Bulletin of the New York Academy of Medicine. 51(11):1235.

Wells JC, DeSilva JM, Stock JT (2012). The obstetric dilemma: An ancient game of Russian roulette, or a variable dilemma sensitive to ecology? American journal of physical anthropology. 149(S55):40-71.

Wheeler DA, Srinivasan M, Egholm M, Shen Y, Chen L et al. (2008). The complete genome of an individual by massively parallel DNA sequencing. Nature. 452(7189):872-876.

Wiedenheft B, Sternberg SH, Doudna JA (2012). RNA-guided genetic silencing systems in bacteria and archaea. Nature. 482(7385):331-338.

Wilson D, Wilson E (2007). Rethinking the Theoretical Foundations of Sociobiology. The Quarterly Review of Biology. 82(4):327-348.

Wilson EO (2012). The social conquest of earth. Liveright.

Wittman AB, Wall LL (2007). The evolutionary origins of obstructed labor: bipedalism, encephalization, and the human obstetric dilemma. Obstetrical & gynecological survey. 62(11): 739-748.

Wolynes PG, Eaton WA, Fersht AR (2012). Chemical physics of protein folding. Proceedings of the National Academy of Sciences USA. 109(44):17770-17771.

Wrangham R, Conklin-Brittain N (2003). Cooking as a biological trait. Comparative Biochemistry and Physiology-Part A: Molecular & Integrative Physiology. 136(1):35-46.

Xing C, Qiao H, Li Y, Ke X, Zhang Z, Zhang B, Tang J (2012). Fractal Self-Assembly of Single-Stranded DNA on Hydrophobic Self-Assembled Monolayers. The Journal of Physical Chemistry B. 116(38):11594-11599.

Yuan J, Yankner BA (2000). Apoptosis in the nervous system. Nature. 407(6805):802-809.

Zaccone P, Cooke A (2013). Vaccine against autoimmune disease: can helminths or their products provide a therapy? Current Opinion in Immunology. En Imprenta.

Zhang D, Kilian KA (2013). The effect of mesenchymal stem cell shape on the maintenance of multipotency. Biomaterials. En Imprenta.

CONCLUSION: THE HUMAN FUTURE

"The reason why the universe is eternal is that it does not live for itself; it gives life to others as it transforms"

— Lao Tzu

Sex and the Superorganism

Gene shuffling to produce variability is pervasive, and as pointed *ut supra* regarding fractality, acts at different levels of complexity. It has been demonstrated that sexually reproducing species survive longer than those with asexual reproduction, with the exception, naturally, of fast reproducing microorganisms. So, what natural forces make sexual reproduction to appear, evolve and spread throughout nature when, evolutionarily speaking, sex seems to be disadvantageous for the individual? Mating requires excessive energy expenditure in both finding a mate and in the act of copulation itself; increases the risk of being caught by predators; facilitates the spreading of sexually transmitted diseases, and each individual passes only half of its genetic material to offspring. At the individual level sexual reproduction does not make too much sense, but indeed, the ubiquitous nature of sexuality tells us a whole different story.

Particularly, a theoretical model based on negative epistasis has strongly called the attention of evolutionary biologists and geneticists since it could explain the evolution of genetic recombination during meiosis (Kondrashov, 1988; 1995). A deleterious allele is less harmful when it is alone than when accompanied by other deleterious alleles in the rest of the genome. The negative effects of this type of alleles are exacerbated —when they are together— in a non-additive but mutually reinforced manner. Recombination greatly helps putting these alleles together resulting in poorly adapted phenotypes, thus, facilitating their elimination.

What's wrong with this model within evolution toward complexity? Although negative epistasis could explain the long term survival of sexually reproducing species and other evolutionary conundrums, the strong negative selection does not explain how sexuality contributes to the creation of complexity since its only function is to purge deleterious alleles. But indeed, sexual reproduction must have been an important element in evolution to complexity since sexual species are mainly the ones that have evolved towards more complex stages (recall that asexual bacteria failed to jump into the matazoarian world). Therefore, it should be a strong mechanism of contribution to complexity, but by other means.

Central to the theory of evolution toward complexity as presented here are two main factors: A few positive changes that contribute to enhancement and specialization of individuals, and extensive simplification —much of which might be caused by deleterious and other not-so-advantageous mutations —leading to interdependence. Sexuality should contribute to the creation of complexity allowing the survival of a high number of genetic variants. For example, diploidy and random segregation of chromosomes during meiosis may allow the survival of deleterious alleles whose function is compensated by the normal ones (resesivity). Additionally, haplotypes containing deleterious alleles can be separated through meiotic recombination, allowing survival of the bad alleles through neutralization. Evidence is compatible with the fact that both positive and negative mutations are essential in the evolution of sexuality (Hartfield and Keightley, 2012; Jiang et al., 2013). However, neutralizing effects of sexual reproduction over "not so good" alleles may contribute the most to complexity in humans and other eusocial mammals. Consequently sexuality should aid in group selection allowing the survival of a greater number of allelic variants, maintaining a genetic pool that can be used at any moment. During evolution of complexity this pool can be use as a source of variability, individual divergence, simplification, and specialization, all of which could contribute to create interdependence, division of labor, cooperation, and thus, adaptation at the group level.

And so the things, a question arises: ¿Is there any evidence of genetic influence in division of labor within human complex societies? Well, at least some (Nicolaou and Shane, 2010). Genetic differentiation among human subgroups inhabiting a territory arises from lack of sexual encounters between those subgroups, which might be a function of distance or sexual preclusion. In a genetically subdivided population, individuals mate preferentially with others inside the same subgroup, thus, differentiating from the others (Holsinger and Weir, 2009; Bamshad et al., 2003; Basu et al., 2003). Factors leading to subdivision include race, ethnicity, socioeconomic background, and religion. It is not far-fetched the idea that within a big, structured city, many generations of marriages between people closely related by their professions could generate some differentiation towards "specialized" occupational genes.

However, cells forming multicelular organisms are characterized for being genetically homogeneous. If we assume that human populations are evolving toward superorganisms, we must also assume that genetic homogenization will slowly take place in them. Metazoarians usually develop from a single cell to a complex organism composed of many cells, with division of labor. Differentiation within the organism takes place by means of epigenetic changes (usually gene methilation) in genomes, and not by somatic mutations that change genetic sequences. In the same manner, ones the superorganism is fully formed, individuals within it should be highly homogeneous, genetically speaking.

Although some occupational substructure will keep forming from relative isolation of human groups with the same occupations, individual epigenetic modifications added to brain plasticity imposed by social environment and education will most probably be the mechanisms toward specialization, division of labor, and further evolution to complexity.

In most metazoarians, only a small fraction of the cells that compose the organism specializes in sexual reproduction and the same stands for communities of eusocial insects, as bees and ants. Reproduction of these superorganisms is still autonomous, i.e., is of a hermaphroditic nature (every colony counts with both sexes), and sexual dimorphism has not yet appeared at this level of complexity, but sexuality is limited to the queen and drones simulating the egg and sperm.

If we continue with the logical sequence of human communities evolving toward superorganisms, we reach an uncomfortable conclusion: reproductive functions should finally be a specialized activity limited only to a fraction of the inhabitants. Reproduction should slowly disappear from the most part of the population. In a genetically homogenous superorganism the generation of variability within it is no longer needed, but it will still be necessary for populating special habitats inside and outside planet Earth. Standard humans, ideal stereotypes, will rise in proportions within populations. Transformation of regular humans into this models, most likely long-living, healthy, athletic, good looking, sexually attractive, extroverted, and intelligent (but not extremely), will be possible thanks to genetic editing, progress in aesthetic surgery, tissue engineering, and other technological aids. Under this conditions, sexual reproduction as a mean of gene shuffling will not be too necessary and could experience evolutionary relaxation. Of course, this does not signify the end of sex. We would not allow that to happen!

Is there any proof supporting the claim that environmental pressure driving maintenance of sexual reproduction in human populations is somewhat relaxing? I think the answer is yes. For instance, demographic transition, i.e., the transition from high to low birth and death rates as a society develops from a pre-industrial to an industrialized economic system has being amply demonstrated, with a decrease in the average number of children per family through the decades of the postwar, mainly in European nations (Galor and Weil, 2000; Galor, 2005). Quality of sperm counts in men worldwide has steadily dropped in the last century as suggested by several studies (Merzenich et al., 2010), and sexual orientation toward homosexuality and sexual diversity has risen in developed countries since gay liberation movements during the 60s and the 70s (Adler, 2013). These manifestations of reduction of sexuality as a plain tool for reproduction may be product of the transition of cities from simple communities of humans to true superorganisms.

We have traveled a long road from the first sedentary communities constructed alongside rivers, to modern mega-cities. But the growth of a city may have a limit imposed by not yet known constraints, and accumulation of complexity into higher levels has been boosted recently. It is likely that this trend will be exacerbated in the next decades.

Globalization and open markets

Let's go back to intelligence, and thus to the steep road. Recall communication through chemicals in simple multicellular organisms, the formation of specialized structures for transportation, and the emergence of primitive nervous systems forming networks. The appearance of nervous systems in the evolutionary arena allowed the formation of larger and more complex multicellular organisms, but life had to wait hundreds of millions of years until the emergence of a true, advanced intelligence.

No doubt the overall human superorganism is at an early evolutionary stage. We jumped from chemical, visual and auditory communication to electrical networks like mobile telephones and internet in just a couple of centuries. But these networks are still just that: relatively simple, flat networks. We do not observe in those networks the type of complexity found in central nervous systems of higher animals, with sophisticated mechanisms of local and global control, or specialized hierarchical structures. Human brains, for example, are the most complex structures known to us, harboring an extraordinary high number of connections, organized in several levels of complexity (Saver, 2006). Nevertheless, we can observe how the Global Superorganism starts taking life on its own, and with it, the emergence of new global problems.

In the nineteen nineties, in its march to the Global Superorganism, humanity plunged into a new evolving global trade network called globalization (He and Deem, 2010). Markets were liberated and borders for economic transactions began to disappear (Bekaert et al., 2003). Around the world, peasants began to place their products in the global market and could know the price to sell at all times thanks to an internet connection. Traders, businessmen, industrialists and financiers from around the world saw their transactions and their seeking for strategic partners greatly streamlined thanks to email. Globalization and liberalization of the markets brought previously unseen phenomena, such as the unprecedented economic growth in emerging economies of Asia, which was called "the Asian miracle" (Stiglitz, 1996; Nelson and Pack, 1999). New technologies and methods used by the Global Superorganism promised to eradicate poverty from the world.

But the instability of the nascent global Superbrain, still immature, consisting of interconnected human beings with access to the global information, internet portals acting as ganglions, and liberated markets as flow of energy between related cooperative subsystems (nations), soon became evident. The explosions of speculative economic bubbles began to affect not only local economies, but spread epidemically from one country to another. In the globalized world, a recession triggers a loss of confidence that travels over the network as a ripple, without recognizing borders, stalling the economy, producing inflation and unemployment, and making the payment of huge national debts impossible. Since the beginning of the popularization of the Internet about two decades ago, the world has experienced several recessions of regional or global scale. These recessions are expression of self-organized criticality at a global level (Stiglitz, 2000; Ormerod and Heineike, 2009; Imbs, 2010).

Moreover, not only the legally established economies have globalized, but also those that are part of organized crime. International drug traffic and other criminal groups are becoming powerful transnational corporations carrying violence to many countries (Morselli, 2011). In addition, genetic, cultural and educational differences between people within and among populations, have determined the emergence of individuals that fit extremely well into the era of rapid changes of the global economy, allowing them to astronomically increase their fortunes in the overnight aggravating the wealth disparity which, as it has been demonstrated, is the basic cause of most social problems (Wilkinson and Pickett, 2011). These problems are not new, but globalization has given them a new impetus.

The Global Superorganism

Besides global economy and malignancy, the global Superbrain has given a major proof of its existence. Perhaps there is no better signal of the rise of the global Superconsciousness as its own defense against attempts to silencing it. In 2011, US representatives proposed a bill designed to expand law enforcement for internet companies that violate copyrights. Although it was a local, national law, the global Superorganism interpreted SOPA (Stop On-line Piracy Act) as a serious threat to its freedom and existence, and through major neural centers like Google, Wikipedia and other websites, an unprecedented self-defense world-wide protest was organized. Pharmaceutical companies, media businesses, the Motion Picture Association of America and other SOPA supporters as well as promoter representatives, were targeted for a boycott that included denial of service attacks and innumerable petition drives from US voters and citizens from all the countries of the world. Other defense mechanisms included blackouts and an opposition rally held in Ney York City (Sell, 2013). It is no more a matter of personal opinion; it is a matter of the harmonic response of the Global Superorganism. It already has its own agenda. No need to say SOPA never saw the light.

In addition to development of more potent and sophisticated self-defense mechanisms, it is likely that the future global brain will gradually evolve to dampen and control economic instabilities, acquiring resilience and diminishing the potential global impact derived from the collapse of any local economy. It is also possible that the new emerging collective intelligence will offer creative solutions to help minimize global social inequality, the scourge of global crime and other global diseases.

The real future global superbrain will continue to evolve in complexity to automatically control all transactions between nations. In the future, every metropolis (or state depending on the case) of the globalized world will specialize in the production of one or a few items that will export to the rest of the globe, increasing interdependence among nations. As cell, cities will be autonomous entities with respect to simple, housekeeping activities, but at the same time will become increasingly simplified as the number of specialized production niches within them decreases. The overall superbrain will regulate the amounts produced to exactly meet demand, that is, in real time but making long-term projections; will adjust economic systems controlling or freeing prices according to the convenience of the markets and the people; will make early corrections to avoid recessions and extreme inequality between citizens; will build the confidence derived from the exact knowledge of the values of the economic parameters in real time and their future estimates, and will automatically switch from models of austerity and economic downturn to models of aggressive investment and high indebtedness and vice versa, as convenience.

The huge Global Superbrain will not only control the world economy, but all kinds of parameters. It will allow real-time participation of citizens and governments in managing issues of various kinds (Lee, 2013), from scientific to political—as for instance, will replace representative democracy for a much more participative one—channeling the collective intelligence. Through various types of devices carried by people, vehicles, and installed in all parts of the world as if it were a colossal sensorial system, it will assess global climate, monitor public health and take action in case of epidemics, watch the skies for asteroids and other threats, plan and control the global air traffic, monitor water sources, control the power supply of the cities, and so much more. It will handle all aspects of humanity in which we currently consume a lot of personal intelligence, time and other important resources, but will also be able to make creative decisions. Of course, the raise of the Global Superbrain will carry with it new instabilities and externalities; new diseases of the Global Superorganism, as it happens with each level of complexity.

It's hard to imagine the Global Superbrain in further stages of evolution, and all the fantastic emergent properties it would present. It is likely that it will develop a type of self-awareness and personality located well above our personal human scales of space and time. We would not perceive that higher intelligence, or communicate with it, at least in the same way in which we communicate among us. It is also possible that it will have a sort of free will. Although we will be integral (but tiny) parts of that overall intelligence, it would transcend us in the same way that our self transcends each of the neurons that make up our brain.

Who knows how long will it take for the network to evolve into this sort of supermind, but one thing is certain: a brain is meaningless without interaction with other brains. By the times when humans have conquered other planets, we will be talking about communities of planets. Perhaps, we will have discovered entirely new forms of communication (maybe faster than light eliminating the paradigm of relativistic restriction described in Einstein, 1995) and faster transpobrtation. We should not worry too much about the technical problems. The planning and construction of spacecrafts will be swift little details for the powerful terrestrial brain.

When will this trend toward complexity stop? It will not stop. If we do not succeed; if we humans finally end up destroying our own environment, and with it, ourselves; if we were doomed to autoannihilation, sooner or later another species will rise. Is life against death, is creation against destruction, is the ability living systems have to lush and expand indefinitely when the necessary time and adequate energy resources are provided, against destruction and final annihilation: the heat death of the universe.

But then again, it is possible that the seed of life had been planted in other planets, perhaps many. At least in some of them civilization might have arisen and with it, interplanetary travel.

The meeting of two interplanetary civilizations could be a catastrophic event, or maybe because of high intellectual development they might have mastered evolution to complexity, learned to appreciate life, and preferred to engage in cooperation eventually deriving, with time, in symbiosis. The encounter of two intelligent worlds might originate something new, with new emergent properties. But who knows? It is slightly probable that such an encounter had already taken place in our Earth and modern society really emerged with the help of a race of giant space travelers.

And then at last, in a distant future, after many eons had passed and life had sown bacteria, complex societies, and planetary superorganisms in every nook and cranny of the cosmos, then, and only then will the universe come to live. That will be the time of the Universal Superorganism.

<h2 style="text-align:center">LITERATURE CITED</h2>

Adler MA (2013). The ALA Task Force on Gay Liberation: Effecting Change in Naming and Classification of GLBTQ Subjects. Advances in Classification Research Online, 23(1):1-4.

Bamshad MJ, Wooding S, Watkins WS, Ostler CT, Batzer MA et al.(2003). Human population genetic structure and inference of group membership. The American Journal of Human Genetics. 72(3):578-589.

Basu A, Mukherjee N, Roy S, Sengupta S, Banerjee S et al.(2003). Ethnic India: a genomic view, with special reference to peopling and structure. Genome research. 13(10):2277-2290.

Bekaert G, Harvey CR, Lundblad CT (2003). Equity market liberalization in emerging markets. Journal of Financial Research. 26(3):275-299.

Einstein A (1905). On the electrodynamics of moving bodies. Annalen der Physik. 17(891):50.

Galor O (2005). The demographic transition and the emergence of sustained economic growth. Journal of the European Economic Association. 3(2–3):494–504.

Galor O, Weil DN (2000). Population, technology, and growth: From Malthusian stagnation to the demographic transition and beyond. The American Economic Review. 90(4):806–828.

Hartfield M, Keightley PD (2012). Current hypotheses for the evolution of sex and recombination. Integrative Zoology. 7(2):192-209.

He J, Deem MW (2010). Structure and response in the world trade network. Physical review letters. 105(19):198701.

Holsinger KE, Weir BS (2009). Genetics in geographically structured populations: defining, estimating and interpreting FST. Nature Reviews Genetics. 10(9):639-650.

Imbs J (2010). The first global recession in decades. IMF economic review. 58(2):327-354.

Jiang X, Hu S, Xu Q, Chang Y, Tao S (2013). Relative effects of segregation and recombination on the evolution of sex in finite diploid populations. Heredity. In Print

Kondrashov AS (1988). Deleterious mutations and the evolution of sexual reproduction. Nature. 336:435–440.

Lee N (2013). E-Government and E-Activism. In: Facebook Nation (pp. 115-146). New York: Springer.

Lovelock J (2003). Gaia: the living Earth. Nature. 426(6968): 769-770.

Merzenich H, Zeeb H, Blettner M (2010). Decreasing sperm quality: a global problem?. BMC Public Health. 10(1):24.

Morselli C, Turcotte M, Tenti V (2011). The mobility of criminal groups. Global Crime. 12(3):165-188.

Nelson RR, Pack H (1999). The Asian miracle and modern growth theory. The Economic Journal. 109(457):416-436.

Nicolaou N, Shane S (2010). Entrepreneurship and occupational choice: Genetic and environmental influences. Journal of Economic Behavior & Organization. 76(1):3-14.

Ormerod P, Heineike A (2009). Global recessions as a cascade phenomenon with interacting agents. Journal of Economic Interaction and Coordination. 4(1):15-26.

Saver JL (2006). Time is brain — quantified. Stroke. 37(1):263-266.

Sell SK (2013). Revenge of the "Nerds": Collective Action against Intellectual Property Maximalism in the Global Information Age. International Studies Review. En Imprenta.

Stiglitz JE (1996). Some lessons from the East Asian miracle. The World Bank research observer. 11(2):151-177.

Wilkinson RG, Pickett K (2011). The spirit level. Bloomsbury Press.

ABOUT THE AUTHOR

Dr. Edwin F. Herrera-Paz is a physician and human geneticist who studies ethnic populations in Honduras, Central America. He is a professor of human genetics and physiology at the *Universidad Católica de Honduras* and international consultant to CAAPA (Consortium on Asthma in African-ancestry Populations of the Americas), a project leaded by Johns Hopkins University at Baltimore, Maryland. His work has been published in prestigious peer reviewed scientific journals.

www.ingramcontent.com/pod-product-compliance
Lightning Source LLC
Chambersburg PA
CBHW021406170526
45164CB00002B/532